Metrics for
Elementary and
Middle Schools

Metrics for Elementary and Middle Schools

By V. Ray Kurtz

The Curriculum Series

nea

National Education Association
Washington, D.C.

Library of Congress Cataloging in Publication Data

Kurtz, V. Ray.
   Metrics for elementary and middle schools.

   (The Curriculum series)
   Bibliography: p.
   1. Metric   system—Study   and   teaching
(Elementary)—United States. 2. Metric system—
Study and teaching (Secondary)—United States. I.
Title. II. Series.
QC97.K87   389'.152   77-27973
ISBN 0-8106-1714-5-00 Paper
ISBN 0-8106-1715-3-00 Cloth

CONTENTS

The Author
Dr. V. Ray Kurtz is Professor of Curriculum and Instruction in
the College of Education, Kansas State University, Manhattan.
He is author of another book on metrics, *Teaching Metric Awareness*,
and for several years he was a public school teacher in Kansas.

The Consultants
The following educators have reviewed the manuscript and provided
helpful comments and suggestions: Joyce Eaton, remedial mathematics
teacher, Hempstead School, St. Louis, Missouri; Genevieve M. Ebbert,
elementary teacher, Maria Hastings School, Lexington, Massachusetts;
Mary Ann Jackman, mathematics laboratory teacher-coordinator,
William M. Trotter School, Boston, Massachusetts; Susan Duraski
Kramme, sixth grade teacher, Ellisville Intermediate School, Ellisville,
Missouri; and Jo Anne Schaack, third grade teacher and mathematics
adviser, Plainview Avenue Elementary School, Tujunga, California.

INTRODUCTION

Metrics for Elementary and Middle Schools is designed for teachers of these grades who want a practical down-to-earth explanation of those problems and questions concerning the metric system which are bothering everyone. Such questions as Why are we changing? What is wrong with the old system? and How can I become prepared to teach the new system? will be answered through the format of explanation and metric activities.

This book is divided into three parts. The first part is a primer of information designed to help a teacher in answering the many questions that children and parents will ask. This part also includes practice problems for all teachers involving the mathematics of metrics. The second part consists of ideas and activities that teachers of the nonmathematics curriculum may incorporate into their lessons to aid in the total metric experience of the learner. The third part deals with classroom activities that teachers of mathematics may use. A special section is included which uses only nonstandard measurement activities designed for primary children. Attention is also given in this section to the use of metric measurements in presenting activities that are concrete, semi-concrete, and abstract in nature. Regardless of the age of the learner, if an understanding of the system is not present, the starting place is with the concrete weighing, pouring, and measuring activities. Such an approach is used in this book.

Metrics should be taught to children as a primary system. Conversions from the English system to metrics should not be used or encouraged. Conversion tables which are given in Appendices B and C are informational in nature and are not intended for use by students. The ultimate goal of metrication is for the student, parent or teacher to be able to conceptualize the metric units and be able to work mathematically with them so that when a gram, liter, or centimeter is discussed, the learner will understand its size. This awareness of metric units can come only from actual experiences with measurements. These experiences must be of a concrete nature.

The ability to work mathematically with abstract metric problems must be acquired after the student is familiar with the various metric units of length, weight, area, volume, and temperature. This book is designed so that abstract activities will follow the concrete and semi-concrete activities.

In conclusion, I would like to have you think about a poem written by my wife—

A current issue confronting our nation
Causing great exasperation,
Fear, disdain, and consternation
Is that of total metrication.

The '75 Conversion Act
Was passed by Congress to, in fact,
Effect a change quite permanent
In the U.S. system of measurement.

With the coming of metrics to the U.S.A.
Inches and pounds become passe,
Yards and feet give way to meters,
Quarts and pints are measured in liters.

As we change our ways to measure and weigh,
Dissident folks view this change with dismay.
All the new methods are bound to seem strange.
There will be people who don't want to change.

A good attitude toward the metric way
Can be fostered by teachers at work and play.
In setting examples you must agree
The classroom teachers hold the key.

METRIC PRIMER FOR TEACHERS

GENERAL INFORMATION

Why Go Metric? This question is being asked over and over by persons in the United States. The main problem is that many persons who ask the question do not listen with an open mind to the answers. The author has found students and teachers to be much more receptive to the new system than some members of the general public. It is quite common to find persons so much against the new system that they say, "Why should the United States adopt the measurement system of the world? Why doesn't the world adopt the measurement system of the United States?" Some persons go so far as to say it is a communist plot to have one worldwide measurement system. It is quite unlikely that such persons will voluntarily accept the new system. This does not mean, however, that the new system will fade away. It only means that these persons will be out of step with the remainder of the populace.

The metric system is decimally based, which makes it much easier to use than the customary system. If it is desired to change from one metric unit to another metric unit, the decimal point is moved to the right or to the left, for all units of the same kind are increased or decreased either by multiplying the standard unit by a power of 10, or by dividing by a power of 10. Which is easier to change—3.341 miles to feet or 3.342 kilometers to meters? The first problem requires multiplying by 5,280. The second problem requires moving the decimal point three places to the right.

The second reason for going metric is that the metric system is similar to our money system and our numeration system. Actually if a learner understands money and place value, there is very little more that needs to be learned. These relationships should be stressed to children. A child who understands place value will have little trouble understanding the metric system if the similarities between the two systems are stressed, as one concept can be built upon another. The systems dovetail. The similarities are presented in depth in a later section.

The third reason is that the rest of the world is metric or going metric. It is difficult to go alone in the world. It has been estimated that the United States will achieve several billion dollars more in world trade after going metric. The question has been asked if a citizen of a foreign country will buy a machine that metric wrenches will not fit.

The fourth reason is the relationships of the new system. All units are linked together, e.g., the cubic decimeter has a capacity of one liter which will hold one kilogram of water. The system has numerous interrelated features. The basic relationships are shown in the following table:

Volume	Capacity	Weight of water
cm³	mL	g
dm³	L	kg
m³	kL	metric ton

A less important but still strong argument for changing to the metric system is that the change will provide an opportunity to clean house in some manufacturing areas. Garment manufacturers see metric conversion as an opportunity to reduce the number of standard clothing sizes. Machine manufacturers are seizing the opportunity to reduce the number of standard sizes of nuts, bolts, screws, and other fasteners.

Objections to Changing. There is always comfort and security in staying with the old and familiar. Conversely, it is a threatening situation to discard the old for the new. This principle obviously applies during those discussions of metric change when it is common for tempers to flare and normally rational persons to become quite irrational.

The major objection to metric conversion has been that the changeover costs will be great. In education there will be the expense of changing textbooks, retraining teachers, and numerous other costs such as purchasing films, teaching aids, etc. In business and manufacturing, high cost is again the primary reason given for not converting to metric. The key to lowered costs, however, appears to be gradual change rather than sudden change whether in textbooks or new model cars.

Industry has found that a sudden shift to metric is very costly. However, a gradual shift as new models are phased in actually costs little more than normal spending for redesigning and retooling. The following statement of Everett Baugh, General Motors official in charge of metric planning, is typical of such exaggerated original cost predictions.

> When General Motors first analyzed the cost of going metric, in 1966, it was about as staggering as the national debt. In 1972, we restudied the cost based on going metric only with new models as they came out. That figure was just 28 per cent of the first estimate. In 1973, another study brought it down to 19 per cent. Now, practical experience suggests that the real cost will be only 4 per cent of the original estimate.[1]

Disconcerting Metrics and Some Answers. Anything new always seems to have inconsistencies and difficulties. The metric system is not without such problems which cluster around several topics. The first we will discuss is spelling. Some persons are exerting considerable energy arguing over how we should spell the basic unit of length. Their question is, "Shall we spell it *meter* or *metre?*" The lines were drawn early in the present metrication attempt and the battle continues to rage. The professionals seem to be the ones most caught up in the struggle. The average person couldn't care less. He or she is still trying to conceptualize the unit, let alone spell it. The same problem is evident in the spelling of the basic unit of capacity. Should it be *liter* or *litre?*

A simplified explanation seems to be that those who prefer *meter* believe this is more consistent with present spelling of similar words. Those who choose *metre* claim this is the world spelling and we should "adopt not adapt." Additionally, they say that the United States is flanked on the north and south by Canada and Mexico who are adopting the *re* spelling; England and the European continent are also using *re*. So we come to the moment of truth, "How shall I tell my students to spell the terms?" You might as well be honest and tell them that both spellings are used. If your textbook includes metrics, you probably can't go wrong by using its spelling. You may want to set a good example by not using an inordinate amount of energy in fighting this battle. Move to more fertile fields by providing the students with many metric activities of a concrete, semi-concrete, and abstract nature.

You may also see the symbol for *dekameter* as *dkm* rather than *dam*. This situation does not present as big a difficulty, as the latter spelling is more appropriate.

Another disconcerting problem is the use of the term *weight* or *mass* when reading the results when you step on the bathroom scales. This time the difficulty arises because *weight* is actually in error unless you are at the equator, at sea level, or in a vacuum. Under these conditions mass and weight are numerically equal. As you move away from these conditions, however, they do not remain numerically equal. Mass is a universal unit and remains constant no matter where it is located. Weight varies according to its location because it is dependent upon the pull of gravity which varies from one location to another. The mass of an object on the moon and on the earth would be the same. The weight of an object on the moon would be only approximately one-sixth of its weight on earth because the pull of gravity is much less on the moon than on earth. Therefore, a person who weighed 60 kilograms at the earth's equator would weigh only 10 kilograms on the moon. The mass, however, would measure 60 kilograms on earth and on the moon. Even though this difference in mass and weight exists, the numerical amount on earth varies very little. The 60 kilogram person would weigh only slightly less on top of the highest mountain in America.

The mass-weight dilemma may be solved by the following guidelines. In technical and scientific work a clear distinction should be made between the two terms. In everyday use, the word *weight* nearly always means *mass* and is, therefore, acceptable. If you are interested, the unit for measuring weight or the pull of gravity is the *newton*. As you are aware, this unit is not commonly used. Therefore, it seems proper to go along with the common labeling procedures that we see on foods, e.g., weight: 293 grams.

Measurement of the Past. In 1791, the French General Assembly adopted the concept of a system of weights and measures founded on the meter. The meter was defined as one ten-millionth of the distance from the north pole to the equator along the meridian near Dunkirk in France and Barcelona in Spain. Area and volume were then related to the square meter and cubic meter and weight to a standard metric volume of water.

Confusion followed the adoption of this system. In 1840 the French passed a law forbidding the use of any system other than metric. This second law marked the beginning of French leadership in metrics. All the farsighted leaders of that earlier period were not found in France alone, however. Thomas Jefferson proposed in 1790 that the United States go metric. Needless to say, his proposal failed.

After the metric system became compulsory in France, it was widely used and spread rapidly to other nations. In 1866, the United States Congress enacted legislation authorizing the use of metric weights and measures in all contracts, dealings, or court proceedings. Even though the system never achieved widespread general use in the United States, it was used extensively in scientific and technical work.

While France was developing the metric system, America was using a hodgepodge of measurements known as the customary or English system of weights and measures. This system, which included Anglo-Saxon, Roman, Babylonian, Egyptian, and Nor-

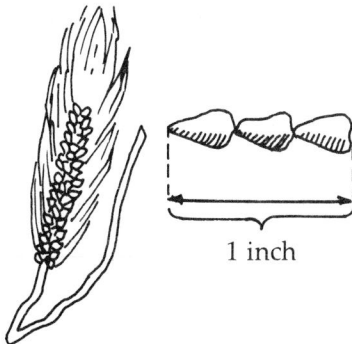

1 inch

man French weights and measures, was primarily based on physical measures. For example, originally the foot was the actual length of a human foot. Often the standard became the king's foot. In order to improve measuring procedures which relied on the human body, other familiar physical standards were used. In 1305, King Edward I of England decreed that three grains of barley, dry, round, and taken from the middle of the ear, when laid end to end should equal one inch. The yard as decreed by King Henry I was the distance from the tip of his nose to the end of his thumb on his outstretched arm.

Measurement based on physical lengths varied greatly. Enterprising businessmen soon learned to have a long-armed person buy the cloth and a short-armed person sell it. Even though the standards were not good, relationships between units in the system were worse. The units were not related to one another in a predictable way, e.g., there were 5,280 feet in a mile, 12 inches in one foot, and 16 ounces in a pound. These units are not linked together by an overall simple rule as in the metric system; therefore, when compared to the metric system, the customary system is considerably more cumbersome.

A Danish astronomer, Ole Rømer, in 1702 first made a reproducible thermometer using the melting point of ice and the boiling point of water as fixed points, and dividing the scale into equal increments. This method remains in use today.

A young instrument maker, Daniel Gabriel Fahrenheit, visiting Rømer in 1708, observed how the old scientist calibrated thermometers. After several years of experimenting,

Fahrenheit, in 1717, began making mercury thermometers commercially. His final scale was based on two fixed points, the melting point of ice (32°) and the heat of the healthy human body, commonly called blood heat (96°). Fahrenheit stated that the boiling point of water was 212°, but did not place such a high temperature on the thermometer. Since the correct body heat is 98.6° F, it was probably fortuitous that he arrived at the correct boiling point of water. Soon after his death, the boiling point of water replaced blood heat as the upper fixed point on the thermometer.[2]

Although the scales used on early thermometers varied greatly, apparently no one thought of using 100 degrees between the melting point of snow and the boiling point of water. Rømer had 60° as the boiling point of water and 7½° as the melting point of ice. The Fahrenheit scale, of course, needs no explanation.

It is common knowledge that Anders Celsius had a thermometer with 100 degrees between the boiling point of water and the melting point of snow on December 25, 1741. Celsius, a professor of astronomy at Uppsala, Sweden, from 1730 until his death in 1744, was very active in the early thermometer experiments. What is not commonly known, however, is that his scale was inverted from our thermometers of today. It read 100° for the melting point of snow and 0° for the boiling point of water. It was not until after the death of Celsius in 1744 that the thermometer was reversed to its present form.

Even though available history does not build an ironclad case that Celsius was the first to come up with a centigrade thermometer, he was certainly among the first. Because his name is most often associated with the centigrade temperature scale, the Ninth General Conference of Weights and Measures in 1948 officially ruled that what had been called degrees centigrade should henceforth be called degrees Celsius. It is appropriate that one who pioneered in this area should receive such recognition.

If you enjoy history at its best, be sure and read the book by Middleton cited earlier, especially pages 66-105, for further information on this subject.

Measurement of the Present and Future. Today, virtually the whole world is metric or going metric. This country started moving officially toward becoming a metric nation when President Ford signed the Metric Conversion Act of 1975. Even though the act was very permissive, it gave teachers and others involved with education justification for moving rapidly toward the new system.

England started the changeover in 1965 but made little progress during the first ten years. The country had just changed to a decimally based money system and people thought businessmen had taken advantage of them. They were in no mood to make another change and be taken advantage of again. Additionally, a dual system has been used which stressed both methods. Retaining both systems does not appear to be the most efficient way to change a nation to metrics.

Canada waited for the United States to go metric, finally gave up in 1973, and started on its own. Since that time they have made excellent progress. They seem to be relying less on a dual system than did England. An example of the singular Canadian approach—the government weather bureau made an early start by giving the temperature only in degrees Celsius. Such a procedure forces the listener to become familiar with the new system since there is no opportunity for the temperature to be given in Fahrenheit. When a dual system is used, probably little progress is gained other than familiarity with terms, as in reporting the temperature on bank signs in both Celsius and Fahrenheit. It is encouraging that the United States Weather Bureau plans to use only metric starting June 1, 1980, for reporting temperature, rainfall, and wind velocity. This single move by the Weather Bureau will be a giant step toward encouraging the citizenry to become familiar with the metric system of measurement. The weather forecast of the future might be: "The high for the day will be 31°C, the low will be 25°C, the wind will be out of the north with gusts up to 30 kilometers per hour." Since people love to talk about the weather, they will be forced to use the new system.

Goal of Metrication. Some trite sayings have emerged in the metric movement which actually turn some people off. Even with the possibility of losing a few readers, there is really no better way to express the goal of metrication than the saying, "Think Metric." This platitude means that the student should think in terms of metric units rather than in customary units. You as an adult are likely to continue thinking in terms of the traditional units; some practicing teachers may not achieve metric fluency. If someone were to ask you to estimate the height of a person and you thought first that there were 2.54 centimeters in an inch and the person was about 65 inches tall, you would be dragging yourself through the traditional system to reach the metric answer.

The goal for children is to learn the metric system of measurement as their primary system. If they "spoke metric," they would think directly in the metric system. In the preceding problem a child thinking in metric would say, "The person is about 165 centimeters tall." Naturally this is an acquired skill that does not happen overnight. With many actual measuring activities, however, children are able to think metric as their basic measurement system. There may be some children in the changeover phase who become able to think in either measurement system. Just the mention of a particular measurement would trigger them to shift into the appropriate system. They would not need to translate back and forth as would a person with only one measurement language. It is not likely that there will be large numbers of these "bimeasural" persons just as there are not large numbers of bilingual persons.

The changeover in children's textbooks appears to be coming quickly. As this book is being written, textbooks are emerging that stress only metric problems. The authors of these texts feel that since the nation has fixed its destiny by passing the Metric Conversion Act of 1975, we might as well get the children changed over as quickly as possible. These children will be able to pick up in an incidental way all that they need for using the customary system. Remember, no teacher is doing a young child a favor by helping him or her learn to think primarily in the traditional system at the exclusion of the metric system. A child so equipped will be at a disadvantage in the world in which he or she will live.

Home and Industry Responsibilities. Even though teachers have a great responsibility to roll up their sleeves and get busy helping to provide children with basic metric understandings, others also

have equally important roles in a smooth changeover. Young children will be terribly confused if they are faced with metric units in one class and customary units in another. Possibly even more confusing for a child would be to work with the metric system at school and then see and hear nothing but feet, pounds, and inches at home, on TV, and in the stores. Dr. Floyd David, director of the Office of Education's Metric Education Programs, stated this dilemma very well, "Teachers who switch to international units may find they are beating their heads against a stonewall if their students have to deal in pounds, feet and quarts everywhere but in school."[3]

Le Systeme International d'Unites (SI). Since the French General Assembly adopted the concept of a system of weights and measures founded on the meter in 1791, there have been many additions and alterations to the system. The modernized system, called *SI,* was established in 1960 by international agreement. It should be noted that SI is different in a number of ways from former practices. At one time *cc* was an acceptable abbreviation for cubic centimeter. Presently the correct symbol[4] is *cm*³. Additionally, the degree centigrade has changed to the degree Celsius as a tribute to the man who founded the thermometer with 100 degrees between the freezing and boiling points of water. British and American spellings of some metric units (e.g., *metre* or *meter,* and *litre* or *liter*) also differ. These spelling differences appear to be consistent with other word variations in the two countries.

The Department of Commerce issued a notice in the December 10, 1976 *Federal Register* updating metric style and usage for the United States. These recommendations included the following: "The international symbol for liter is the lowercase 'l' which can easily be confused with the numeral '1'. Accordingly, the symbol 'L' is recommended for United States use." The Department of Commerce interpretations included the *er* spellings in *meter* and *liter* and the *da* symbol for the prefix *deka.* They also stated, "Weight is the commonly used term for mass."[5]

These recommendations are consistent with those made by the American National Metric Council which favor the *er* spellings, the symbol *L* for *liter, da* for the prefix *deka,* and the term *weight* for everyday use. They recommend dropping the comma to separate groups of digits. The digits in a numeral should be separated into groups of three, counting both to the left and to the right of the decimal point. A space the same width as that formerly occupied by the comma should be placed between the groups of digits.[6]

Examples:[7]   342 497.243 7
0.423 73

The recommendations of these two agencies are believed to be forward-looking; they are, therefore, incorporated into this book.

Before specific units are presented, the metric framework of prefixes will be explained. The overriding beauty of metrics is the consistent, clear pattern that pervades the system. Instead of repeatedly writing the expressions for 100 times the unit, 10 times the unit or 1/10 times the unit, we adopt the prefixes *hecto, deka,* and *deci.* The scheme is efficient and easily learned. These prefixes are used with the various metric units whether of length, weight, or capacity. The more commonly used prefixes follow.[8]

Prefix	Symbol	Pronunciation
mega	M	as in *mega*phone
kilo	k	as in *kilo*watt
hecto	h	heck'toe
deka	da	deck'a (*a* as in *a*bout)
deci	d	as in *deci*mal
centi	c	as in *senti*ment
milli	m	as in *mili*tary
micro	μ	as in *micro*phone

A dilemma awaits each teacher concerning the units to be taught. The choice is between teaching the structure and relationships of the system or teaching what is practical and useful. The advantage of stressing each position will be presented.

Teaching Structure. There has been considerable agreement among educators during recent years that if the learner can see a pattern of relationships among concepts, learning is easier and is not as readily forgotten. This principle applies strongly to learning the metric units. If the similarity between money, place value, and the metric system is to be stressed to the fullest, all units need to be presented. Following are the three decimally based areas for metric length, money and place value.

Metric	kilometer	hectometer	dekameter	meter	decimeter	centimeter	millimeter
Money	$1,000	$100	$10	$1	$.10	$.01	$.001
Place Value	thousands	hundreds	tens	ones	tenths	hundredths	thousandths

Some educators feel that omitting the little used length unit *hectometer* is as serious as leaving out hundreds in place value. If the hectometer is left out, the pattern is broken and the total picture is difficult to see.

Teaching the Practical and Useful. Educators who stress teaching only the most frequently used units believe that teaching all units unnecessarily complicates the system and causes students to become overwhelmed by the large amount of material. The units of each quantity that are used most frequently in the practical, everyday world follow.

Most Commonly Used Units

Quantity	Unit	Symbol
Length	kilometer meter centimeter millimeter	km m cm mm
Weight	metric ton (1000 kg) kilogram gram milligram	t kg g mg
Area	square kilometer hectare* (dam^2) square meter square centimeter	km^2 ha m^2 cm^2
Volume	cubic meter cubic decimeter cubic centimeter	m^3 dm^3 cm^3
Capacity	liter milliliter	L mL

*Accent the first syllable. The second syllable should rhyme with *care*.

Most Commonly Used Units—*continued*

Quantity	Unit	Symbol
Time	day hour minute second	d h min s
Temperature	degree Celsius	°C

The author believes that the structure approach should be stressed with children in the initial stages of learning the metric system. Showing the direct similarities between metric, money and place value has strong pedagogical value. It is a good learning practice to demonstrate to a child that the metric system is structurally no different from two other situations with which he or she is familiar. The student should not be too upset to learn later that a hectometer is used very little once he or she gains an understanding of the metric patterns.

Comparison of Practical Units. Even though comparisons between units in the customary system and the metric system are to be discouraged with children, it may be helpful for adults and older students to compare the two systems. Some of the following comparisons are made with metric units and some with customary units.

How will we measure length?
a. By using the millimeter (mm). A millimeter is about the thickness of a dime.
b. By using the centimeter (cm). A centimeter is about the width of a woman's small fingernail.
c. By using the meter (m). A meter is slightly longer than a yard.

How will we measure capacity?*
a. By using the milliliter (mL). Five milliliters are equal to one teaspoon.
b. By using the liter (L). The liter is a little more than a quart.

How will we measure volume?**

---

*The amount a container will hold is called capacity.
**The amount of space inside a container is called volume.

a. By using the cubic centimeter (cm³). The cubic centimeter holds one milliliter.
b. By using the cubic decimeter (dm³). The cubic decimeter holds a liter.
c. By using the cubic meter (m³). The cubic meter holds a kiloliter. Full of water, it would weigh a metric ton.

How will we measure weight?
a. By the gram (g). A regular M&M candy weighs about a gram. A dollar bill weights a gram. A nickel weighs five grams.
b. By the kilogram (kg). A kilogram is slightly heavier than two pounds.

How will we measure temperature?
By the degree Celsius.* One degree Celsius (°C) is approximately twice as large as one degree Fahrenheit.

Metric Scope and Sequence. Since the United States has decided to go metric, there is no reason to spend much time discussing with students and colleagues the advantages and disadvantages of the new system. The main questions now center around scope and sequence, or (1) How much? and (2) When?

Even though children vary considerably in their ability to work in a meaningful way with measurement, guidelines should be established on *when* to teach specific measurement concepts. Richard Copeland, who not only has interviewed Jean Piaget, but has made an extensive study of his research, states:

> . . . If systematic measurement is to be "taught" it should not be presented before the latter part of what is usually the third grade. Even then, for most children it will have to be an experimental or trial-and-error readiness-type experience. The necessary concepts, as usual, develop from within rather than without for operational understanding. You cannot tell children how to measure . . . they should be allowed to experiment and try to solve measurement problems for themselves. The teacher should play the role of questioner in moving toward the objective desired. The necessary concepts will develop (1) when the child is old enough (eight and one-half, according to Piaget), and (2) when he is allowed to operate on (experiment with, manipulate) objects used in measurement. Both conditions are necessary for the operational thought necessary to perform measurement.[9]

*The Celsius scale is identical to the centigrade scale. Common usage prefers Celsius over centigrade.

In preparing for implementation of the metric system in Hawaii, King and Whitman described the outcomes of their research: "Our findings are consistent with those of Piaget and suggest that our students do not possess the concepts necessary for understanding measurement until sometime during the third grade."[10]

They decided that for kindergarten through second grade most measurement activities would be trial-and-error readiness-type in order to allow the child to develop measurement and pre-measurement concepts using nonstandard units. Instruction in formal measurement using standard units would be emphasized in third grade and an understanding of measurement would be achieved by the end of sixth grade.

This author feels some uneasiness in suggesting that kindergarten, first, and second grade teachers not become involved in teaching standard measures such as centimeter and meter, because he is aware of specific situations where teachers on this level have had very rewarding metric units. After taking an in-service course many primary teachers become very enthusiastic about the metric system and want to jump in and help their children master the new system. This feeling is laudable. It seems reckless, however, to suggest going against the available research. Since children vary greatly in their readiness to pursue academic topics, each teacher must use his or her professional judgment as to when individual students are ready for the transition to measuring the length of the room in centimeters from hands, pencils, etc.

An observant teacher can readily determine when a child is ready to measure with standard units of centimeters and meters. In first or second grade the teacher might lead a discussion as to whether or not the two opposite walls of the classroom appear to be the same length. The teacher might then suggest that the class measure the two walls using a nonstandard unit, such as shoe lengths. After carefully selecting two children with shoes of different lengths, the teacher might say, "Bill, use your shoe and measure the wall. Susan, use your shoe and measure the wall." When the measurements are completed, most children will accept the fact that one wall is 35 shoes long and the other 41 shoes long. However, the members of the group who are uneasy and question what is wrong are the ones who are ready to use the more formal measuring of the standard units of length of the metric system. For your convenience, the activities of this book are divided into (1) nonstandard measurement activities and (2) standard measurement activities.

Accompanying the change to the metric system are two scope and sequence difficulties: an earlier need for decimals and negative numbers. Since the metric system is a decimal system of measurement with all units divided in tens, there is much less need for fractions such as fourths, eighths, twelfths, thirty-seconds, etc., and more need for decimals. Despite this change of emphasis, no educators are advocating the elimination of fractions from the curriculum. Many educators now agree, however, that decimals should be offered at an earlier age with the introduction of beginning concepts in grade four. In general, fractions and decimals will change places in the sequence of the mathematics curriculum in the elementary school. It is anticipated that as the metric change nears completion, and the need declines for more complex fractions (such as fourths, eighths, sixteenths and sixty-fourths), the time spent on such fractions will be greatly reduced.

Some teachers will anticipate this change with a great deal of enthusiasm while others will be displeased and not understand the lesser importance that fractions will eventually possess. The coming of the metric system may be less of a shock for some teachers than the realization that decimals rather than fractions will have become the backbone of rational numbers.

The earlier need for negative numbers will dictate that integers be taught sooner in a child's educational career. The main need comes in using the Celsius scale where all temperatures below freezing must be expressed in negative numbers. Teachers of a strong science program have long been aware of an earlier need for a knowledge of negative numbers.

The Use of Diagnosis. As metric emphasis gets underway, it will become necessary to diagnose a child's knowledge of the metric system. It is true that regardless of grade level, if he or she is unable to conceptualize metric measures, that child needs more concrete experiences with those units. It is also true that if he or she can conceptualize the units, that child does not need more and more concrete experiences with only slight variations. For the next several years the degree of metric emphasis from classroom to classroom will vary considerably. It behooves the conscientious teacher, therefore, to find out exactly what the child learned about the metric system before coming to a new class. This diagnosis can be made from a pencil and paper test where students are diagnosed on their knowledge of the system. A diagnosis could also be obtained by asking questions orally such as "What is the teacher's height in centimeters?" If the child responded with 20 cm or 200 cm, you would know measuring activities were needed.

MATHEMATICS OF METRICS

This section is designed to facilitate the teacher's task in reviewing or learning for the first time the mathematics involved in working with metric units. In no way should this presentation even hint that the abstract, pencil and paper activities should precede concrete experiences with the metric units. Since adults are often impatient learners wanting to learn the new material right now, this abstract section is placed at this location in the book. Remember, you want to gain (1) an understanding of the metric units and (2) an ability to change from one metric unit to another. This section is designed to aid you with the second goal.

After working these problems, you will agree that the mathematics involved is not difficult. The person who understands how to multiply and divide by moving the decimal point will agree that the metric system is the easiest measurement system ever devised.

The first step will be to review the metric prefixes. The six most often used prefixes, their values, their symbols and the unit are as follows:

Prefix	kilo	hecto	deka	unit	deci	centi	milli
Value	1000	100	10		0.1	0.01	0.001
Symbol	k	h	da		d	c	m

These are universal prefixes and may be used with *meter, gram,* or *liter.* In every case the prefix *kilo* is 1,000 times the unit, i.e., a *kilogram* is equal to 1000 *grams.*

If you want to change 434 g to kilograms, you must decide how many places and which direction to move the decimal point. If you refer to the preceding table, you will notice that kilo is three places to the left of the unit which would represent grams. Therefore, you would move the decimal three places to the left. The answer would be 0.434 kg. In changing from a larger unit to a smaller unit, you move the decimal point the opposite direction. For example, to change 3.65 kg to grams, you would move the decimal three places to the right. The answer would be 3650 grams. Some additional problems follow. Remember, when going from smaller units to larger units, move the decimal to the left, and when going from larger units to smaller units, move the decimal point to the right.

Problems: (Answers will be found on pages 28 and 29.)

1. a.	1 kg	=____g		m.	29 cg	=____dag
b.	2.3 kg	=____g		n.	17 mg	=____g
c.	0.234 kg	=____g		o.	24 cg	=____dag
d.	3000 g	=____kg		p.	342 mg	=____g
e.	4291 g	=____kg		q.	42 dag	=____g
f.	321 g	=____dag		r.	34 hg	=____dag
g.	421 g	=____hg		s.	321 dg	=____hg
h.	7.4 kg	=____g		t.	7634 mg	=____dag
i.	3 g	=____mg		u.	632 mg	=____hg
j.	3.62 dg	=____mg		v.	32 dag	=____cg
k.	63 g	=____mg		w.	4.9 hg	=____dg
l.	7 cg	=____mg		x.	7.3 kg	=____mg

Next you will find some fill-in questions related to linear measures. The problems are worked exactly like the preceding ones on weight. In problem 2.a. fill in the blanks with numerical values equal to 4000 meters.

2. a.	_4000_ m	=____ km	=____ hm	=____ dam			
b.	____ km	=____ m	=_300_ dam	=____ hm			
c.	_0.3_ km	=____ dam	=____ m	=____ hm			
d.	____ m	=____ km	=____ hm	=_3.9_ dam			
e.	____ dm	=____ m	=_7.3_ cm	=____ mm			
f.	____ m	=_734_ mm	=____ cm	=____ dam			
g.	____ mm	=_43.6_ cm	=____ m	=____ dam			
h.	_3.5_ dm	=____ cm	=____ m	=____ dam			
i.	____ m	=____ cm	=_3.924_ dm	=____ hm			
j.	_0.063_ km	=____ hm	=____ m	=____ dm			
k.	____ m	=_0.035_ dam	=____ hm	=____ km			
l.	____ mm	=____ dm	=____ m	=_0.0039_ km			
m.	____ mm	=____ m	=_0.349_ dm	=____ cm			

n. _____ dm   =_____ m   =_____ cm   = 0.555 mm

o. _____ m   =_____ dm   = 37.42 hm   =_____ km

Some of the following problems may cause you to turn to the appendices for conversions. If you can work these correctly, you are well on your way to being a metric expert.

3. Convert:
    a. 750 grams to kilograms
    b. 472 milligrams to grams
    c. 75 grams to milligrams
    d. 3.2 kilograms to grams
    e. 730 kilograms to metric tons
    f. 29.3 metric tons to kilograms
4. Add:
    a. 7.32 kilograms + 490 grams. Give answer in kilograms.
    b. 34 kilograms + 4963 grams. Give answer in kilograms.
    c. 250 kilograms + 7.1 metric tons. Give answer in metric tons.
5. Answer:
    a. Are 1000 mL more or less than a quart?
    b. Is a mile more or less than a km?
    c. Are 500 g more or less than a pound?
    d. If candy sells for 95 cents a pound, what should be the selling price for one kilogram?
    e. What is the equivalent speed in kilometers per hour for 55 miles per hour? (Round to nearest 10)

The relationships that exist in the metric system between volume, capacity, and weight are beautiful. However, from teaching many metric in-service classes and workshops, the author is aware that some teachers find these relationships difficult to conceive. You will have a special appreciation for the metric system if you master the following section.

Metric Relationships

Volume	Capacity	Weight of Water
1 cm³	mL	1 g
1 dm³	L	1000 g
1 m³	kL	1 metric ton

Using these relationships, work the following problems.

6. a. If a glass contains 25 mL of water, how many grams of water would this weigh?
   b. If a bucket has a volume of 450 cm³, what would be the weight of the water if the bucket were filled? Give answer in kilograms.
   c. Another bucket weighs 150 grams when empty. When *filled* with water, its total weight is 550 grams. What is the volume of the container in cubic centimeters?
   d. A swimming pool is 5 meters wide, 12 meters long and 1.5 meters deep. When the pool is full, what is the weight of the water in the pool? Give answer in metric tons.

Answers:

1.	a.	1000	i.	3000	q.	420
	b.	2300	j.	362	r.	340
	c.	234	k.	63 000	s.	0.321
	d.	3	l.	70	t.	0.7634
	e.	4.291	m.	0.029	u.	0.006 32
	f.	32.1	n.	0.017	v.	32 000
	g.	4.21	o.	0.024	w.	4900
	h.	7400	p.	0.342	x.	7 300 000

2.	a.	4000	4	40	400
	b.	3	3000	300	30
	c.	0.3	30	300	3
	d.	39	0.039	0.39	3.9
	e.	0.73	0.073	7.3	73
	f.	0.734	734	73.4	0.0734
	g.	436	43.6	0.436	0.0436
	h.	3.4	34	0.34	0.034
	i.	0.3924	39.24	3.924	0.003 924
	j.	0.063	0.63	63	630
	k.	0.35	0.035	0.0035	0.000 35
	l.	3900	39	3.9	0.0039

m.	34.9	0.0349	0.349	3.49
n.	0.005 55	0.000 555	0.0555	0.555
o.	3742	37 420	37.42	3.742

3. a. 0.750 kg
    b. 0.472 g
    c. 75 000 mg
    d. 3200 g
    e. 0.730 metric tons
    f. 29 300 kg

4. a. 7.810 kg
    b. 38.963 kg
    c. 7.350 metric tons

5. a. more
    b. more
    c. more
    d. $2.09
    e. 90 km per hour

6. a. 25 g
    b. 0.450 kg
    c. 400 cm³
    d. 90 metric tons

METRICS FOR THE
NONMATHEMATICS CURRICULUM

No curriculum change of modern times is of the magnitude of the transition from the customary system to the metric system of weights and measures. The modern mathematics change of two decades ago was considered a revolution. The magnitude of that change, however, cannot be compared to the magnitude of the present change. If you weren't a teacher, child or parent, you probably escaped most of the modern mathematics revolution. Uncles, aunts, grandmas and grandpas never needed to become involved with modern math.

Metric change will be different because ultimately all citizens must become familiar with the new system. If they do not, they will be out of step with the remainder of the nation. Some areas which have changed or are planning to change are weather reporting, grain exporting, medicine and drugs, athletics (especially track events), bottling of liquor, manufacturing of automobiles and hardware fasteners, and photography.

This nation is presently in the very early stages of metrication. Changing the citizens from customary to metric will take place over many years. Even though gradual, it appears that changing the adult population is going to be very difficult, especially among those members of society who do not favor it. One commonly hears older persons say, "I hope I am gone when the metric system comes into use." These people are not so much opposed to the country going metric as they are to themselves being forced to convert.

It will be considerably easier to introduce the youth to the use of metrics as their primary system of measurement if they receive help and encouragement at school in all areas of the curriculum. Certainly they may be receiving little help out of school. Hopefully, it will not be left to mathematics and science teachers alone to shoulder the entire load of teaching the new system.

The purpose of this chapter is to present specific ways for teachers of the nonmathematics curriculum to aid in the task. Whether you are a teacher in a departmentalized situation or in a self-contained classroom, the same procedures will apply.

Educators have known for many years that specific applications enhance transfer. Teachers must help students identify facts that can be woven into meaningful generalizations and then provide for the application of the generalizations in a range of situations. The author's goal is to offer ideas and suggestions to teachers of the nonmathematics curriculum so that they can provide the metric applications that will be so desperately needed. It should be noted that transfer is further enhanced when the applications encountered by the student outside of school are similar to those used for instructional purposes. Since the school can only *hope* for metric experiences outside of school, however, the main thrust in school must be for learners to see various applications as they move from class to class or from subject to subject. Too frequently teachers teach metrics in mathematics and science but do not mention it in any other subjects. When this happens, many meaningful applications are overlooked.

The following rock-bottom, basic principle should be kept in mind by all teachers as they contribute to the metric changeover: be positive about the new system, build it up; don't bad mouth metrics. Learners are going to hear many derogatory remarks about the metric system during out of school hours; it will be unfortunate if one of their teachers contributes just one word of opposition. Any teacher should be able to follow this principle. It requires no study of the new system, merely a realization that metric is here and is not going away.

Teachers who are in a departmentalized organization may decide to present a metric unit that cuts across two or more disciplines. The various combinations of disciplines will not be mentioned here. One example with which the author is familiar, however, involved two sixth grade teachers who worked together in preparing their students for the measurement of temperature in degrees Celsius. The children involved in this project functioned mainly in a self-contained classroom but moved to a departmentalized area to study science. The problem encountered by the science teacher was that sixth grade learners had difficulty reading temperature scales which involved negative numbers. Since the science activities required that students be able to add and subtract readings involving integers, preliminary activities were necessary to provide this preparation. The math teacher prepared the learners for the experience by working with ordered pairs of numbers which led to a natural introduction of such ordered pairs as 0 and 32, 100 and 212. The objective of the assignment was for the learners to be able to explain the relationship between the Celsius and Fahrenheit scales without the use of a formula. Both teachers involved with this integrated unit spoke highly of its success. Even though it can be argued that the two scales should not be compared, students as old as sixth graders have had considerable contact with the Fahrenheit scale and can therefore gain meaningful relationships by considering the two scales. The goal remaining is for students to use the Celsius scale as their primary measurement scale.

The following discussion will be concerned with metrics and such curriculum areas as language arts-reading, social studies, music, and art. Science is not included since many metric activities in this area are presently available.

Language Arts—Reading

Those who teach language arts and reading overlook numerous occasions every day to include metric experiences. These opportunities are missed if a conscious effort isn't made to stress metric vocabulary. One problem that will be with us for some time is that the stories in basal readers and trade books will change very slowly to include metric vocabulary. The trade books will not change until replacement books with metric terms are purchased. Therefore, the reading teacher needs to adapt the present material to include metric words. Any time a customary measurement comes up, the teacher can relate it to a metric measure. If the story of "Danny the Dinosaur," for example, states that Danny weighs 8 tons and is 60 feet long, the teacher can casually state that if Danny were here today, we would weigh him in metric tons or kilograms and measure him in meters. Depending on the age of the children, an exact quantitative conversion to metric units may be in order. At most grade levels there are children who are capable of making these conversions. The hand-held calculator could be used for solving problems of this nature.

The problem of obsolete measurement terms found in stories is not unique to the present; it is common for a learner to read literature of the past and run across a statement such as "The farmer plowed a furlong." This term is not understood by learners of today and must be converted to a more familiar expression, for who knows that 8 furlongs are equal to 1 mile? The teacher is the person to provide the bridge between historical and current terms. It may seem unbelievable, but the yard, pound, and gallon will eventually become vestiges of the past, while meter, kilogram and liter will become contemporary measurement units.

Language arts experiences of primary-age learners can include contact with metric measures in the same natural setting as the calendar work used by many teachers at the beginning of each day. Weather reporting can provide such an opportunity with the placement outside the classroom window of a can to act as a rain gauge and a thermometer to record temperatures. These two weather instruments will serve to measure precipitation in centimeters and temperature in degrees Celsius. Such information should be recorded on a weather chart or on the calendar. Teachers of very young learners need to record data themselves during the early part of the year, but should encourage students to do the recording later in the year.

Language arts-reading teachers very effectively use "chart stories" with small groups in writing the learners' own stories describing actual happenings of the day. These stories can easily be adapted to use metric vocabulary. A story could describe a girl named Maria who walked one kilometer to school. It might tell of her little dog that weighed 9 kilograms. Other metric terms could be used to complete the chart story called "Maria's Metric World." The teacher would first place the vocabulary on the board or chart before writing the story as dictated by the children.

The following story was written by a group of sixth graders after they had worked with a metric unit in mathematics. They developed it in language arts for completion by another group of learners in the classroom. The metric vocabulary precedes the story. Words may be used more than once.

METRIC STORY

Vocabulary:

Celsius	milliliters	kilograms
millimeter	meters	gram
kilometers	liter	centimeters
		milligrams

Jack lives 6 (1)_____ from school so he rides the school bus. One morning he noticed that the speedometer read 90 (2)_____ per hour when they were on the highway and 35 (3)_____ per hour in a school zone. Jack was wearing a coat as the temperature was 5° (4)_____. He was glad to be well from the cold he had last week. His mother had given him 325 (5)_____ of aspirin for it.

When he got to school Jack learned that it was the day for the school nurse to measure and weigh the class. Jack was pleased to see that he had gained 2 (6)_____ and was 4 (7)_____ taller than a year ago. It must be because he drinks a (8)_____ of milk every day.

At recess some of the children played football and Jack ran 10 (9)_____ for a touchdown. When Jack went to the cafeteria for lunch he drank 250 (10)_____ of milk with his meal and then ate several pieces of M&M candy which weighed a (11)_____ each.

After lunch Jack's class went to the activity room to see a 16 (12)_____ film. At the end of the day Jack rode home on the school bus and ran in the house to play with his little puppy which weighs 10 (13)_____. It had been a busy day!

Answers:

1. kilometers	6. kilograms	11. gram
2. kilometers	7. centimeters	12. millimeter
3. kilometers	8. liter	13. kilograms
4. Celsius	9. meters	
5. milligrams	10. milliliters	

The following maze adapts well to a spelling unit on metric words. It could be used to introduce the unit.

DIRECTIONS: Circle the metric words in the metric maze and cross them off the word list. The words are written horizontally, vertically, and diagonally. They are all spelled forward. In the diagonal words, the first letter starts at the top and the word is written downhill. Some words overlap and some letters are used more than once. The following metric words may be found in the maze.

temperature	length	degree
ten	metric	volume
meter	hectare	linear
perimeter	dekaliter	celsius
base	kilogram	thermometer
measurement	minute	gram
centigram	standard	capacity
second	ruler	hour
cube	decimeter	system
scale	centiliter	milliliter
liter	calorie	unit
		area

```
T E M P E R A T U R E D M L P A E G B C
E M A T E R F E E D R H E B E B B A S E
N H M E A S U R E M E N T A R N R H D N
A H O L D U M E R S H O E D I H G F M T
M E T R I C X E R T D G R A M O E T M I
I C R D E K A L I T E R F I E M D D H G
N T S T U A E L D R L E J K T D R L Q R
U A V X Z K I L O G R A M U E E Z U D A
T R K L D K G P Q R S T U A R E A V H M
E E V D W X A Y Z D I E R D S E C O N D
S F E C E N T I L I T E R J K S L A D R
T D R D D G N G D A R L L I N C U B E A
A S G E A D R H O S C W D I O A M N S L
N M A C G A O E D Y H E E L N L A O I I
D D R I W R H L E S V O L U M E N T T T
A O P M O L O P E T A L O S E T A G A E
R U L E R I U S P E T D T O I I H R G R
D L U T H E R M O M E T E R L U P E R J
L O T E O C A P A C I T Y O U R S J K R
M A N R D M I L L I L I T E R M Z X S A
```

Social Studies

Those who teach social studies have untold opportunities to build metric concepts as well as to provide numerous reinforcement activities for previously learned concepts. Social studies is replete with quantitative concepts that need more concept-building emphasis than received in mathematics. It is a mistake to assume that a young learner knows the mathematics concepts necessary to understand social studies.

The study of space or geography present the most obvious opportunities to stress metric measurements since the measurement of maps will be changing from miles to kilometers. Young learners will need to gain the concept of a *kilometer*. There is no better way for a child to learn this concept than for the social studies teacher to take the class for a kilometer walk in an area that has previously been measured. Social studies teachers in the past have had children walk a mile to understand the concept of that unit of measure. Walking the kilometer will not only be more appropriate but will be shorter, as a kilometer is only 0.6 of a mile. Following the walk, the class can discuss locations that are one kilometer apart. The discussion can later consider longer distances such as those from one city to another. This activity will be a natural introduction to map reading and scale drawings. Maps that record distances in miles can easily be changed to kilometers by using a hand-held calculator. Such an activity has proved quite successful with learners in fifth and sixth grades.

Another reference problem might relate to the unit of land area, the *hectare*. When learners in fifth and sixth grades are ready to study area, the *are* can be introduced. One successful teacher relates the are to the size of a large garden. Obviously the hectare would be explained as 100 ares. The social studies teacher is in an enviable position to provide concrete concept-building activities relating to the hectare when this term is used in the textbook. Again, walking around a hectare to gain an understanding of its size is a strong concept-building activity.

A third opportunity for metric concept-building is during a discussion of imports and exports, for the metric *ton* has become the world standard for trade. Even in the United States, grain imports and exports are given solely in metric tons. The author has never heard one word of criticism directed toward this policy. All indications point toward total metrication of the grain trade.

For concept-building of the metric ton, a cubic meter may be assembled with twelve metersticks or dowels taped together. If the cube were covered and filled with water, it would weigh one metric ton. This activity has helped many students, young and old alike, to conceive the size of a metric ton.

Music

How many songs do you know that use metric terms? Music teachers have told the author that they have looked in their music books and files and found none. Hopefully, after reading this short section on music and metrics, you will not only become convinced that it is your responsibility to help with the metric changeover, but that you will use one or more of the metric songs with your young learners.

Today most music is taught by a special music teacher. It is common to find pianos in kindergarten rooms, but they are found mainly in the auditorium area or the music room. The suggestions presented in this section can be used by the special music teacher or the teacher of a self-contained classroom.

Numerous songs in our society use customary units in the lyrics. Examples include: "I Love You a Bushel & a Peck" and "Five

Foot Two, Eyes of Blue." As stated before, however, we do not find lyrics using metric terms in songs of today. When one K-6 music teacher was asked, "How can you assist your learners in the metric changeover?" she replied, "I would have to write a song." She had previously been asked to report on the metric songs available to use with children. She had hoped to find a song combining physical movement with metric vocabulary but found no such song. It was at this point that she decided she would need to write such a song herself. A song of this nature that involves actual measurements with the metric vocabulary is not easy to write. The songs that follow were written by teachers who teach young learners every day of the school week.

The first song is designed to combine physical movement with metric terms. Any music teacher can use this song after he or she has learned the physical lengths of a millimeter, centimeter, decimeter, and meter. The song is ideal for second graders, but can be used with both younger and older children. It is sung to the tune of "The Old Gray Mare."

Lyrics written by and used with permission of Mary Jo Kurtz. Sung to the tune of "The Old Gray Mare."

The metric system's here! You'd better learn it now.
Here! You'd better learn it now. Here! You'd better learn it now. The
Metric system's here! You'd better learn it now. Watch us, we'll show you how.
Watch us, we'll show you how. Watch us, we'll show you how. The
Metric system's here! You'd better learn it now. Watch us, we'll show you how.

Can you show a teeny-weeny millimeter?
Teeny-weeny millimeter? Teeny-weeny millimeter?
Can you show a teeny-weeny millimeter? Watch us, we'll show you how.
Watch us, we'll show you how. Watch us, we'll show you how.
Can you show a teeny-weeny millimeter? Watch us, we'll show you how.

Can you show a very little centimeter?
Very little centimeter? Very little centimeter?
Can you show a very little centimeter? Watch us, we'll show you how.
Watch us, we'll show you how. Watch us, we'll show you how.
Can you show a very little centimeter? Watch us, we'll show you how.

Can you show a not-so-little decimeter?
Not-so-little decimeter? Not-so-little decimeter?
Can you show a not-so-little decimeter? Watch us, we'll show you how.
Watch us, we'll show you how. Watch us, we'll show you how.
Can you show a not-so-little decimeter? Watch us, we'll show you how.

Can you take a giant step that shows a meter?
Giant step that shows a meter? Giant step that shows a meter?
Can you take a giant step that shows a meter? Watch us, we'll show you how.
Watch us, we'll show you how. Watch us, we'll show you how.
Can you take a giant step that shows a meter? Watch us, we'll show you how.

The second song was written by a music teacher and a teacher of a self-contained classroom. They reported that a second grade class enjoyed the song a great deal and sang it at a school program. It is appropriate for any age group, including adults.

Lyrics written by and used with permission of Randalin Bargdill and Leola McLain.
Sung to the tune of "Sing A Song of Sixpence."

Sing a song of metrics and have a lot of fun.
Metrics are the "in" thing and fun for everyone.
We can measure lots of things, it's really fun to do. So
Listen very carefully and we will show you, too.

We no longer walk a mile, we walk a kilometer.
When we buy a quart of milk, we buy it by the liter.
Candy is my favorite, I used to buy a pound.
Now I buy it by the gram and pass it all around.

Look at my new pencil, it's seven inches long.
Oh! What did I hear you say, you're really very wrong.
Take your pencil, lay it down beside a meterstick. It's
Eighteen centimeters long; you see there is no trick.

Do you know how much I weigh? I don't mean in pounds.
I don't talk that way now that metrics are around.
I weigh thirty kilograms, that is called my mass.
You may not believe it, but I'm biggest in my class.

We have a new thermometer, it's really very nice. We
Read degrees in Celsius, and zero will make ice. It
Starts right down at zero and goes right on its way.
Twenty-three degrees is sure to be a lovely day.

It's hard to think of big things and just how much they weigh. A
Thousand kilograms makes a metric ton, they say. When
Russia buys our wheat, they buy it by the metric ton.
I know all about it 'cause I'm a farmer's son.

Kilometer, metric ton, gram and centimeter,
Kilogram and centigram, Celsius and a liter.
These are just a few of the words that we can sing. So
Sing a song of metrics, for it is the coming thing.

The third and last song was written by a teacher of a combined fourth and fifth grade. She wrote both the lyrics and the music to be sung by the children in her self-contained classroom while she accompanied on the guitar. She reported that the fourth and fifth graders greatly enjoyed singing the song, and the guitar accompaniment is not difficult.

Lyrics and tune written by and used with permission of Roberta Kinsinger.

1. Change that yardstick to a me - ter, change that quart to a
2. Don't you wor-ry don't you fret, for there's nothing to re-
3. You use ki- lo, hec- to, dek - a, dec - i, cen-ti,mill-i-
4. Then it's ki - lo, hec-to, dek - a, dec - i, cen-ti,mill-i-
5. And it's ki - lo, hec-to, dek - a, dec - i, cen-ti,mill-i-
6. Change that yardstick to a me - ter, change that quart to a
7. Don't you wor-ry don't you fret, for there's nothing to re-

li - ter, and jump on the me - tric wag - on now.
gret. The met-ric sys-tem's ea-sy if you try.
me - ter to mea-sure an-y length that you might find.
gram. They give the weights of eve-ry thing they can.
li - ter — these give-the a - mounts that you de - mand.
li - ter and jump on the me-tric wag - on now.
gret. The met-ric sys-tem's ea-sy if you try.

Art

Those of you who teach art have many opportunities to encourage students to use metric measurements. As one art teacher said, "The secrets are (1) use the metric vocabulary yourself every chance you get and (2) let the students get as much real experience in actual measurement processes as possible." This means to use metric dimensions rather than customary dimensions. If the activity calls for yarn, be sure to give the length in meters. If paint is to be mixed, use cubic centimeters or milliliters. If paper is to be cut, give the dimensions in centimeters and ask the students to do the cutting. If the art lesson calls for margins, give them in centimeters. If you fire a kiln, talk about the temperature in degrees Celsius not degrees Fahrenheit.

A very common art activity is to have the children draw an outline of each other on butcher paper. In order to stress metric length, ask them to measure the outline and to record their height in centimeters on the drawing.

After collecting leaves for the fall leaf design, ask the children to measure the length of the leaves in millimeters and to put the longest leaf in the middle of the design. The length of the longest leaf should be recorded below the design in both centimeters and millimeters.

When giving directions for making a mobile, give the lengths of the cords in centimeters. If it is to be a mobile with geometric shapes extended from a clothes hanger, give the length of the cords as:

15 cm to hold the circle
30 cm to hold the square
45 cm to hold the ellipse
60 cm to hold the triangle
75 cm to hold the rectangle

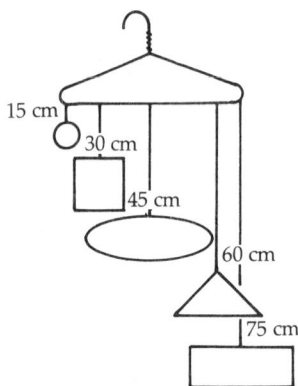

Meter Flower

If the art activity is to make yarn flowers, instruct everyone to measure 1 meter of yarn to use in making the project. Ask the students to glue the petals to the center of a flower as shown in the accompanying diagram. Each petal is made from a 10 cm piece of yarn. The center is a circle made of a 5 cm piece of yarn. The remaining 15 cm of yarn are used in the stem. The centers of the petals may be colored.

The art activity normally called the "magazine picture fold" can be easily altered to use metric measures. A picture from a magazine should be selected and cut into 2 centimeter strips. The strips are then glued back in the same order on a sheet of construction paper, except that a 1 cm space is left between each strip. The effect of the spaces between the strips is interesting.

Paper strip art is readily adapted to metric measurement. Numerous paper strips 2 cm wide and 6 cm long should be cut from various colored construction paper. Each strip should be tightly wound around a pencil and glued. After slipping the strips from the pencil, they may be glued on a paper to form an animal or design. A beautiful three-dimensional butterfly can be made with the curls.

Directions for texture design art can be given utilizing metric weights. If a small design is desired, the directions would be for each student to weigh 5 g each of popcorn, beans, wheat and sunflower seeds. These are to be glued on a 20 cm x 25 cm piece of construction paper forming a design or animal. Other materials may be substituted for those suggested.

Directions for making church windows—a greatly enjoyed activity—can also utilize metric units. Each student should measure and cut two 25 cm x 25 cm pieces of waxed paper. Different colored crayon shavings are then placed between the two sheets of waxed paper and ironed with a moderately hot iron. The resulting product resembles a stained glass window.

Summary

The overall theme that comes through in each subject area is that the teacher should use metric terms whenever measurement terms are needed. In many situations, you, the teacher, can learn the new system along with your students. You should remember that learners need many physical experiences with metric measurements. Most of the activities presented in this section call for actual measurement.

You can assume that teachers of science and mathematics are working to change each learner from being primarily a customary measurer to being primarily a metric measurer. Your role is to encourage and support their work. If the learner's only contact with the metric system is in mathematics and science, it will be a long and difficult task to change him or her to the new system. Each learner needs numerous metric experiences both in and out of school.

You are in a position to talk up the new system as well as to provide reinforcement activities.

MEASUREMENT ACTIVITIES

This chapter includes measurement activities designed for the primary child who needs to develop the concept of measurement as well as activities for the older learner involving more sophisticated measurement. Examples of the former kind, designed for kindergarten, first and second grade children, are included in the first section. In these nonstandard measurement activities the learner does not become involved with standard units of metric measure such as liters, meters, grams, etc. Activities are of a trial-and-error readiness-type which allows time for children to develop measurement and premeasurement concepts using such nonstandard units as finger widths, chalk lengths, pencil lengths, arm spans and strides. Emphasis is placed on the concept of measurement rather than on one correct answer. The exercises are designed to lead into the standard measurement activities which follow.

The second section, metric measurement activities, includes exercises designed as a continuation of the nonstandard measurement activities. They start with concrete measurement at a beginning level. Teachers who lack metric awareness will have no difficulty with the beginning activities of length, weight and capacity. These exercises progress from a very elementary level to a more mature level.

NONSTANDARD MEASUREMENT ACTIVITIES

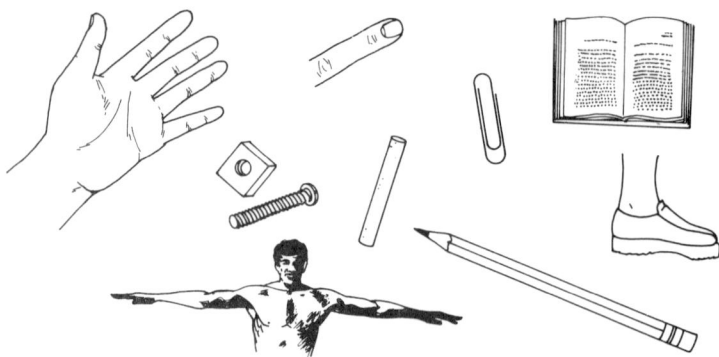

Length

A young learner has a built-in centimeter measure, for the width of his or her finger is approximately one centimeter. The width of an adult's small fingernail also approximates a centimeter. This fact can be used in measuring various objects in finger widths before the concept of a centimeter is discussed.

MATERIALS: Objects to measure such as a new piece of chalk, a new pencil and a book.

NUMBER OF PARTICIPANTS: Any number working in pairs.

DIRECTIONS: Measure the lengths of the objects by using the width of your index finger. After you have completed the measurements, check with other teams to see how your answers agree with their answers.

Length

OBJECTIVE: To measure to determine the order of pictures from shortest to longest.

MATERIALS: Numerous pictures of boats, automobiles, etc., cut from magazines, newspapers, or catalogs.

NUMBER OF PARTICIPANTS: Any number.

DIRECTIONS: Place the pictures in order from shortest to longest. You may want to use the width of your finger to measure the pictures.

Length

OBJECTIVE: To provide a concrete experience using a non-standard unit of length, the finger width.

MATERIALS: Various objects such as paper clips, index cards, sticks of chalk, book, pencil, etc.

NUMBER OF PARTICIPANTS: Any number.

DIRECTIONS: This activity should begin with the teacher explaining how to measure using a finger width. There should be one station for each two children doing the activity. At each station, place one object so that children working in pairs can measure the object as they progress through the stations. After the measurements are made, the teacher should lead a discussion concerning the findings.

Length

OBJECTIVE: To compare the lengths of concrete objects.

MATERIALS: Set of colored rods or strips of paper 1 cm through 10 cm. (No reference will be made to the metric lengths of the rods or strips.)

NUMBER OF PARTICIPANTS: Groups of two or three children.

DIRECTIONS: Provide each learner with the following rods or strips:

Number	Length
1	10 cm
1	9 cm
1	8 cm
1	7 cm
1	6 cm
2	5 cm
2	4 cm
3	3 cm
3	2 cm
1	1 cm

Ask the children to place one of each strip or rod in order from longest to shortest. Ask the children to hold up the longest strip or rod. Hold up one of the 5 cm items and ask how many 5 cm strips or rods it takes to make the longest strip or rod. Vary this procedure with the strips or rods.

Length

OBJECTIVE: To order items according to width.

MATERIALS: Three coins and three cards with arrows.

NUMBER OF PARTICIPANTS: Any number depending upon available materials.

DIRECTIONS:

Place the coins in a triangle as shown. Place the arrows so that each points toward a wider coin.

Length

OBJECTIVE: To compare the height of two learners in developing the concepts of taller, shorter and the same.

MATERIALS: None.

NUMBER OF PARTICIPANTS: Entire class.

DIRECTIONS: Ask two children who are of unequal height to stand back to back in front of the class. Then ask the class to tell you who is taller and who is shorter. Ask two children who are of equal height to stand back to back. Discuss with the class that the two class members are of equal height. Select one-third of the learners and ask them to find a partner who is taller. After this is completed, ask a different one-third of the class to find a partner who is shorter. Finally, ask the remaining one-third to find a partner of equal height.

Length

OBJECTIVE: To gain an understanding of tallest and shortest.

MATERIALS: Butcher paper, newspapers, glue, water-base paints and brushes.

DIRECTIONS: Have children work in pairs tracing each other's outlines on butcher paper or newspapers that are glued together.

These profiles may be painted and cut out. Then they may be put in order to find out the tallest boy, shortest boy, tallest girl, shortest girl, tallest person in class and shortest person in class.

Length

The ability to estimate is a desired skill that can be developed in learners if the appropriate activities are used. Measuring with body spans is such an activity (i.e., the distance that can be reached from tip of finger to tip of finger with arms outstretched). The distance will approximate one meter for young learners. This approach develops estimating skill much better than the use of a meterstick or similar device. Normally the standard unit is not mentioned at least until the activity is completed, after which a discussion may be held concerning meter length.

MATERIALS: None.

NUMBER OF PARTICIPANTS: Any number working in pairs.

DIRECTIONS: Estimate the distance of the back of the room, the side of the room and the length of the chalkboard in body spans. After estimates have been made, make the actual measurements by having your partner count the number of body spans of each dimension. Remember a body span is as far as you can reach with your hands stretched out to the sides.

Length

OBJECTIVE: To provide concrete experience with nonstandard measures while diagnosing which children are ready to use standard measures.

MATERIALS: None.

NUMBER OF PARTICIPANTS: Total class divided into two groups.

DIRECTIONS: After a discussion of the fact that the two opposite walls appear to be the same length, the teacher suggests that both groups measure the walls in shoe lengths. The teacher should pick two children (one from each team) who have different-sized shoes to use in the measurements. After the measurements are completed, a further discussion should be held to determine which children accept the walls to be different lengths. Those who accept the results are not ready for standard measures of centimeters and meters.

Length and Weight

OBJECTIVE: To differentiate between longest, shortest, highest, lowest, and heaviest and lightest.

MATERIALS: Various objects such as toys, boxes, bottles, pencils, etc.

NUMBER OF PARTICIPANTS: Groups of two or three children.

DIRECTIONS: Place several items with each group before asking each group to select the longest item. Repeat the questions until all the words are used. During the exercise, some questions will arise, such as "If we lay the bottle down it becomes the longest item." Questions like these are developmental in nature and have strong pedagogical value. For an evaluation have the children perform the tasks individually as requested by the teacher.

Length

This activity can easily be used as a transitional activity for introducing standard measurements. After completing it, you may want to pose the question, "How would you write to your grandmother and tell her the size of our classroom? Would it be appropriate to use strides to describe the size?" Some students will mention that a standard unit should be used.

MATERIALS: None.

NUMBER OF PARTICIPANTS: Any number.

DIRECTIONS: Ask each child to make a strip of paper as long as his or her normal stride. Suggest that the paper be used to measure the width of the front of the classroom. After this is completed, lay the lengths of paper on the floor from shortest to longest. Discuss why the classroom width varied with the stride of different children.

Weight

The amount of equipment needed for teaching weight activities discourages many teachers from involving students in weight experiences. This difficulty is overcome by using readily available materials. The activity is designed to provide an experience with the concept of weight in an informal approach. The teacher may wish to follow up the lesson with a discussion of standard units. Since the flashlight D battery weighs approximately 100 grams, the objects can be compared to 100 grams.

MATERIALS: Five D batteries, a plastic bag, and numerous objects to be weighed, i.e., chalkboard eraser, rock, book, etc.

NUMBER OF PARTICIPANTS: Any number.

DIRECTIONS: Determine as closely as you can the weights of the objects by holding an object in one hand and as many batteries in the other hand as necessary to balance or equal the weight of the object. If more than one battery is required, you may place them in the bag for more convenient handling.

Weight

OBJECTIVE: To mark the heavier object.

MATERIALS: Sheets of paper showing objects.

NUMBER OF PARTICIPANTS: Any number.

DIRECTIONS: Place an X on the heavier object.

Weight

OBJECTIVE: To provide concrete experiences with nonstandard units of weight.

MATERIALS: Balances and numerous weights such as bolts, nuts, washers, paper clips, etc.

NUMBER OF PARTICIPANTS: Any number working in groups of three or four.

DIRECTIONS: After a discussion of weight, the teacher asks the students to weigh a math book, chalkboard eraser, new stick of chalk, etc. The results should be given as so many bolts, washers, etc. Hopefully some students will suggest that it is difficult to compare the weights of two items if one weighs 3 bolts and the other 14 washers.

Area

OBJECTIVE: To gain the concept that area is measure on a flat surface.

MATERIALS: Newspaper or newsprint, scissors, glue.

NUMBER OF PARTICIPANTS: Any number working in groups of two or three.

DIRECTIONS: Lay a double sheet of newspaper on the floor and ask how many students can stand on the newspaper. After determining the number, cut out shoe prints and paste them on the sheet after painting them. If desired, newspapers may be glued together to make an area of one square meter. The area of the paper may or may not be stressed, depending upon the readiness level of the students.

Area

OBJECTIVE: To measure the area of the bulletin board with concrete nonstandard units.

MATERIALS: Several sheets of construction paper.

NUMBER OF PARTICIPANTS: Groups of two.

DIRECTIONS: After the teacher leads a discussion concerning the meaning of *area,* the area of the bulletin board is measured by pinning sheets of construction paper on the board. If the board doesn't measure evenly in whole sheets, the remaining portion may be estimated or sheets may be folded and cut before being pinned up.

Variation: The teacher could ask what other items might be used to measure the area of the bulletin board. Hopefully other-sized sheets of paper, chalkboard erasers, etc., will be mentioned.

Capacity

In developing the concept of capacity, young learners need experiences with nonstandard units. These activities provide the readiness needed before the study of liters and millimeters is begun.

MATERIALS: Various containers such as coffee cans, milk cartons, jars, etc., and water. One small container such as a frozen juice can is needed as the nonstandard unit of capacity.

NUMBER OF PARTICIPANTS: Any number.

DIRECTIONS: Determine the capacity of each container in number of juice containers. Approximations will be needed since most containers will not hold an exact whole number of juice containers of water. After each container is measured, arrange them in order of capacity.

Capacity

OBJECTIVE: To provide a concrete experience with nonstandard units of capacity.

MATERIALS: Pop bottle caps, plastic cups, or small milk cartons.

NUMBER OF PARTICIPANTS: Entire class, working in groups of two or three.

DIRECTIONS: After a discussion concerning the meaning of capacity, ask the children to determine the capacity of the plastic cups or milk cartons by filling them with bottle caps. Discuss why the answers vary.

Volume

In developing the concept of volume, children need experiences where stress is placed on the concept rather than the mathematical computation. This activity is designed to teach the concept of volume. It should be used with third, fourth and fifth grade students who have not previously developed the concept of volume.

MATERIALS: Small boxes of varying shapes and sizes and many centimeter cubes.

NUMBER OF PARTICIPANTS: Depends only upon the availability of materials.

DIRECTIONS: Fill the boxes with the volume pieces and then count the pieces to find the volume. Do not assign units. If it takes 25 pieces to fill the box, say the volume is 25 volume pieces. Arrange the boxes in order from smallest to largest volume. Children who are at ease with the volume pieces may want to assign the specific unit of cubic centimeters to the answer.

Time

OBJECTIVE: To provide a concrete experience with a non-standard unit of time.

MATERIALS: Small plastic bottle filled with water with a hole punched in the bottom edge.

NUMBER OF PARTICIPANTS: Entire class.

DIRECTIONS: After a discussion concerning time, measure how far members of the class can write their numbers while all the water is running out of the bottle.

METRIC MEASUREMENT ACTIVITIES

"When you cannot measure it, when you cannot express it in numbers—you have scarcely, in your thoughts, advanced to the stage of science, whatever the matter may be."*

Lord Kelvin

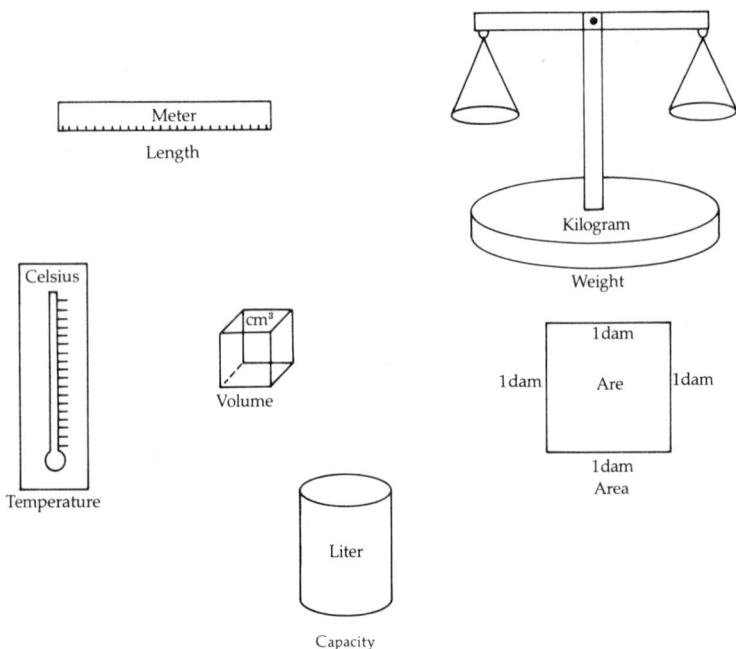

Meter
Length

Celsius
Temperature

cm³
Volume

Liter
Capacity

Kilogram
Weight

1 dam
1 dam   Are   1 dam
1 dam
Area

*This quotation is used by *Science, A Process Approach (SAPA)* to introduce measurement.

Length

Aids for teaching metric length are far more available to the classroom teacher than aids for teaching metric weight, temperature and volume. The author can remember his school days when it was common for a meterstick to be lying in the chalkboard tray. The stick was used by the teacher mainly for pointing. Never was the metric side used for measuring. It was called the meterstick, however, and almost everyone knew that it was a little longer than a yard. This brief contact with metric length describes the major metric experience of many Americans. Needless to say, metric experience of this nature are no longer adequate. This section includes more length activities than activities for any other metric topic because of the availability of length materials and because many teachers introduce the metric system with the subject of length.

The basic unit of metric length is the *meter*. As previously mentioned, the French in the 1790s defined the length of the meter as one ten-millionth of the distance from the north pole to the equator along the earth's meridian running near Dunkirk in France and Barcelona in Spain. In later years, difficulties in reproducing or comparing measurement standards with a portion of the earth's meridian led France to construct a physical standard of the meter of platinum-iridium. Not until 1960 at the General Conference on Weights and Measures was the definition of the meter based on the international prototype of platinum-iridium changed to the following: "The meter is the length equal to 1 650 763.73 wavelengths of the krypton-86 atom in a vacuum."

Length

In order for students to *think* metric, they first must *feel* metric. Many activities involving measurement of linear distances must be planned. These activities require little preparation on the part of the teacher. Students should be encouraged to measure many physical objects in and around school. The activity included here is good, however, as every child can measure the lines at the same time. If the assignment is to measure the width of the door, it will take a considerable time span for a class to make the measurements.

MATERIALS: A sheet of directions and lines for each child, metric ruler.

NUMBER OF PARTICIPANTS: Any number.

DIRECTIONS: Measure the following lines to the nearest millimeter. Record the answer on the line in both millimeters and centimeters.

a. ————————————————————————

b. ——————————————————

c. ——————————

d. —————————————————————————

e. ——————————

f. ————————————————————————————

Answers:

a. 80 mm or 8 cm   d. 85 mm or 8.5 cm
b. 60 mm or 6 cm   e. 35 mm or 3.5 cm
c. 30 mm or 3 cm   f. 105 mm or 10.5 cm

Length—Diagnosis

OBJECTIVE: To estimate centimeter lengths.

NUMBER OF PARTICIPANTS: Any number.

DIRECTIONS: Estimate the lengths of the following lines in centimeters.

a. ——————————————

b. ————————————————————

c. ——————————

d. ———————————————————————————

e. ————————

f. ——————————————

DIRECTIONS: Estimate the following:

I am _____ cm tall.

My teacher is _____ cm tall.

My desk is _____ cm long.

Length

OBJECTIVE: To demonstrate that even traditional sayings will eventually become out-of-date.

MATERIALS: The sayings as listed.

NUMBER OF PARTICIPANTS: Any number.

DIRECTIONS: Place the sayings on the chalkboard and ask students to help you change them so that they use metric terms. Some children may wish to just state equivalency units in metric measures while some may want to completely rework the sayings and state them in whole metric units. The latter is the more desirable method.

"Give him an *inch* and he'll take a *mile.*"

"Traffic just *inched* along."

"Walk a country *mile.*"

"He hit the ball a *mile.*"

"A miss is as good as a *mile.*"

"I would walk a million *miles* for one of your smiles."

"I wouldn't touch him with a ten *foot* pole."

"There was a crooked man and he walked a crooked *mile.*"

"A fourth down and *inches* to go."

Length—Cuisenaire Game

Since Cuisenaire rods have metric lengths, they adapt well to the beginning activities of teaching length. This activity uses only the six shortest rods that measure 1, 2, 3, 4, 5 and 6 cm. One set of rods is required for each of the two players. As you observe the learners playing the game, you will see them measuring the rods on the meterstick. This is exactly what the activity is designed to do, i.e., aid the players to become familiar with 1 through 6 cm lengths.

MATERIALS: One box of Cuisenaire rods for each player, one regular cubical die, and a meterstick.

NUMBER OF PARTICIPANTS: Two.

DIRECTIONS: "Play moves in a clockwise direction with each player rolling the die to determine whether he gets to place a 1, 2, 3, 4, 5, or 6 centimeter length along his side of the meterstick. Play continues until a winner has enough rods end to end to exactly equal 1 meter."*

Meterstick

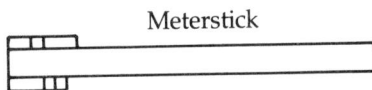

*From Kurtz, V. Ray: *Teaching Metric Awareness* (St. Louis: The C. V. Mosby Co., 1976).

Length—Fill in the Staircase

This activity is designed to accompany the preceding Cuisenaire game. Again, the purpose of the activity is to aid the players in becoming familiar with the 1 through 6 cm lengths.

MATERIALS: One box of Cuisenaire rods and one regular cubical die.

NUMBER OF PARTICIPANTS: Two.

DIRECTIONS: "Each player arranges a staircase of 10 rods. Each then takes a turn rolling the die to determine what length from 1 to 6 centimeters he may use to help form a decimeter square of the staircase. If a player cannot use the designated length, he may not play during that turn. The first player to complete his square wins the game. The staircase and one play is shown in the diagram. Any combination of rods may be used to complete each decimeter length. Rods may only be played vertically."*

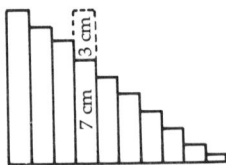

*From Kurtz, V. Ray: *Teaching Metric Awareness* (St. Louis: The C. V. Mosby Co., 1976).

Length

OBJECTIVE: To measure height and reach.

MATERIALS: Meterstick or metric tape.

NUMBER OF PARTICIPANTS: Any number.

DIRECTIONS: Use the metric tape or meterstick to measure your height in centimeters. Measure your reach from fingertip to fingertip. Are they the same? If they are the same, you are a square. Check with your friends to see if they are squares or rectangles. Are there more squares than rectangles in your class?

Length

OBJECTIVE: To use a visual aid when initially presenting the metric prefixes. This aid will certainly gain the attention of the children.

MATERIALS: Eight boxes of various sizes. A tennis ball can or Pringles can may be used as the engine and jello boxes, toothpaste boxes, etc., used for cars. Paper for covering the boxes, magic marker, string or yarn.

DIRECTIONS: Cover the boxes with paper and label them with the proper designations. Join the cars together with yarn. Omit the wheels if they add to the difficulty of the construction project. Use the train in the developmental phase of presenting the metric prefixes.

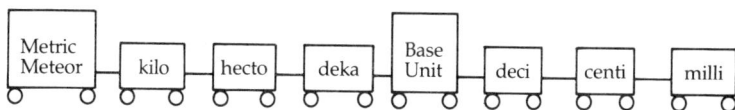

Length—Bingo

Length Bingo is designed to provide a review of the metric terms of length. The game should not be used until class members are able to recognize the words and symbols of length. It is suggested that a seasonal or holiday flavor be added by using appropriate markers such as a bunny, Santa, pumpkin, turkey, etc.

MATERIALS: Bingo cards and key terms as shown.

NUMBER OF PARTICIPANTS: Eight student players and a caller.

DIRECTIONS: After preparing the bingo cards and two sets of key terms, the game is ready to be played. One set of key terms should be cut up and placed in a box and used for calling the terms. These cards may be laminated. The other set is used for calling back terms after someone calls "Bingo."

Seasonal figures may be cut and used for markers on the bingo card or one could be given to each student for the free space. The ways to bingo can be varied, i.e., horizontal, vertical, four corners or blackout.

KEY TERMS

A meter	A m	A metric	A decimeter	A centimeter	A 5 meters
A dm	A cm	A length	A millimeter	A 10 decimeter	A 6 centimeters
B length	B metric	B meter	B decimeter	B dm	B millimeter
B centimeter	B cm	B m	B 10 decimeters	B 6 centimeters	B 5 meters
C centimeter	C cm	C meter	C decimeter	C dm	C m
C millimeter	C 10 decimeters	C 5 meters	C metric	C length	C 6 centimeters

LENGTH BINGO

A	B	C
Length	meter	dm
cm	Think Metric	centimeter
metric	decimeter	millimeter

LENGTH BINGO

A	B	C
length	meter	cm
6 centimeters	Think Metric	centimeter
metric	decimeter	m

LENGTH BINGO

A	B	C
dm	decimeter	length
m	Think Metric	metric
meter	cm	centimeter

LENGTH BINGO

A	B	C
m	metric	millimeter
decimeter	Think Metric	6 centimeters
length	5 meters	cm

LENGTH BINGO

A	B	C
dm	decimeter	length
m	Think Metric	metric
cm	dm	centimeters

LENGTH BINGO

A	B	C
m	length	decimeter
metric	Think Metric	centimeter
dm	millimeter	meter

LENGTH BINGO

A	B	C
5 meters	m	6 centimeters
centimeters	Think Metric	millimeter
millimeter	decimeters	meter

LENGTH BINGO

A	B	C
dm	6 centimeters	5 meters
10 decimeters	Think Metric	m
metric	cm	length

Length

OBJECTIVE: To estimate centimeter lengths.

MATERIALS: Paper, pencil and centimeter ruler.

NUMBER OF PARTICIPANTS: Any number.

DIRECTIONS: Without using a ruler, draw three lines (not in a straight line) so that the sum of the lengths will total 15 centimeters. After you have drawn the three lines, measure them with a ruler to see how close you were to 15 centimeters. Have a contest with your friend to see who can come the closest to the stated length. Change the length on each try.

Length

OBJECTIVE: To measure a 30 m distance and then realize the length of this distance.

MATERIALS: Trundle wheel or meterstick.

NUMBER OF PARTICIPANTS: Any number.

DIRECTIONS: Measure off 30 m on the school ground. Have a classmate take your pulse while you are standing or sitting. Record the results. Next, run a 30 meter dash and have your classmate again take your pulse. Graph the results. Compare your results with other students.

Length

Most of your students will be much more familiar with units of the customary system than with units of the metric system. You can generally assume that they have little prior knowledge of the actual length of a kilometer. You may wish to review that a kilometer is a little over 0.6 mile. This will provide students with an idea of the distance of a kilometer. The best procedure for teaching the concept of a kilometer is to have the learners measure off a kilometer distance.

MATERIALS: Trundle wheel or appropriate tape measure.

NUMBER OF PARTICIPANTS: Teams of two or three members plus a measurement team of two or three members.

DIRECTIONS: Ask the teams to estimate from a beginning point each of the divisions of 100 meters up to and including 1 kilometer. Each team is required to place a mark at each of the 10 points. The team coming closest to the true kilometer distance as measured by the measurement team is declared the winner. Five city blocks are approximately 1 kilometer in length.

Length

OBJECTIVE: To experiment with various pendulum lengths.

MATERIALS: String and a weight.

NUMBER OF PARTICIPANTS: Groups of two or three students.

DIRECTIONS: Measure a 1 m 10 cm length of string. Tie one end of the string to a nail, hook, etc., in a doorway. Tie a weight to the other end so that the length of string between the weight and the hook is 1 m. Draw the weight back and let the weight swing. Count the number of swings per minute. Shorten the length of string between the hook and weight to 50 cm and repeat the experiment. What are your conclusions?

Length

You will find that a short review of the type of decimal problems encountered in metrics will be an excellent introduction to the abstract part of length. This work should involve multiplication and division practice with 10, 100, and 1000.

OBJECTIVE: To provide a review of multiplying by 10, 100, and 1000.

MATERIALS: Problems duplicated from those given and placed on study sheets or the chalkboard.

NUMBER OF PARTICIPANTS: Any number.

DIRECTIONS: Multiply by 10, 100, or 1000 as directed.

1. 8 x 10	8. 5.3 x 1000
2. 90 x 10	9. 5.47 x 100
3. 2.4 x 10	10. 0.632 x 1000
4. 0.36 x 10	11. 0.376 x 100
5. 5.9 x 100	12. 1.346 x 1000
6. 0.63 x 100	13. 0.34 x 1000
7. 0.7 x 100	14. 0.7 x 1000

Answers:

1. 80	8. 5300
2. 900	9. 547
3. 24	10. 632
4. 3.6	11. 37.6
5. 590	12. 1346
6. 63	13. 340
7. 70	14. 700

Length

You will notice that when dividing by 10 or a multiple of 10, the division is very easy. Actually the regular division process will not be used as soon as the learner sees that the decimal point may be moved to the left the same number of places as there are zeros in the divisor.

OBJECTIVE: To provide a review of dividing by 10, 100 and 1000.

MATERIALS: Problems duplicated from those given.

NUMBER OF PARTICIPANTS: Any number.

DIRECTIONS: Divide by 10, 100, or 1000 as directed.

1. 50 ÷ 10	8. 7963 ÷ 1000
2. 51 ÷ 10	9. 94.3 ÷ 100
3. 0.7 ÷ 10	10. 0.7 ÷ 1000
4. 49.2 ÷ 10	11. 52.4 ÷ 100
5. 763.7 ÷ 10	12. 0.763 ÷ 10
6. 7.63 ÷ 100	13. 0.34 ÷ 100
7. 773.5 ÷ 1000	14. 0.9 ÷ 1000

Answers:

1. 5	8. 7.963
2. 5.1	9. 0.943
3. 0.07	10. 0.0007
4. 4.92	11. 0.524
5. 76.37	12. 0.0763
6. 0.0763	13. 0.0034
7. 0.7735	14. 0.0009

Length Rummy

OBJECTIVE: To add metric lengths of meter and kilometer.

MATERIALS: 52 cards as follows:

2 marked 500 m
10 marked 250 m
10 marked 100 m
10 marked 25 m
10 marked 50 cm
10 marked 25 cm

NUMBER OF PARTICIPANTS: Three or four.

DIRECTIONS: 1. Shuffle cards and deal seven to each player. The dealer places the remaining cards face down on the table and turns the top card face up next to the deck to begin the discard pile.

2. Play starts with the player to the left of the dealer. Player may take the top card from the face-down cards, the top card from the discard pile, or the entire discard pile. His or her turn is ended by discarding a card unless he or she has run out of cards.

3. The object is to form and lay out a "book" of cards that equals 1 km or 1 m. Additionally, three cards of a kind form a book.

4. Play continues until a player runs out of cards.

5. To score, a 1 km book is worth 3 points, a 1 m book is worth 2 points, and three of a kind are worth 1 point. In order to encourage taking the entire discard pile, no penalty is assessed for cards held when another player goes out.

Length

The following activities are designed to be worked with the assistance of the hand-held calculator. Students will be unable to work these problems unless they have a thorough grasp of the value of each metric unit of length. This activity is an excellent check on a student's understanding of metric prefixes and place value.

MATERIALS: Hand-held calculator.

NUMBER OF PARTICIPANTS: Any number.

DIRECTIONS: 1. Enter 2249.369 m into a hand-held calculator

add   39 m
subtract   9 mm
add   12 m
subtract   36 cm
subtract   23 hm

What do you have remaining? (0)

2. Enter 1364.112 m into a hand-held calculator

subtract   12 cm
add   36 m
add   1 hm
subtract   1 dm
add   10.8 cm

What do you have remaining? (1500 m)

(More Difficult)

3. Enter 3333.333 m

subtract   3 hm
add   2 km
subtract   3.3 dam
subtract   3.33 dm

What do you have remaining? (5000 m)

Length

The following activity is designed to follow concrete activities
where learners have had the opportunity to experience
millimeter, centimeter, meter and kilometer lengths.

OBJECTIVE: To provide abstract experience with metric length
terms.

MATERIALS: Duplicated sheets of problems as given.

NUMBER OF PARTICIPANTS: Any number.

DIRECTIONS: Write the proper metric equivalent in the blanks
provided.

1. 1 m = _____ cm
2. 1 m = _____ mm
3. 1 km = _____ m
4. 1 cm = _____ mm
5. 1 mm = _____ cm
6. 1 m = _____ km
7. 14 m = _____ cm

8. 170 m = _____ km
9. 1362 m = _____ km
10. 732 cm = _____ m
11. 0.432 km = _____ m
12. 0.93 mm = _____ cm
13. 342 cm = _____ km
14. 463 cm = _____ m

Answers:

1. 100
2. 1000
3. 1000
4. 10
5. 0.1
6. 0.001
7. 1400

8. 0.170
9. 1.362
10. 7.32
11. 432
12. 0.093
13. 0.003 42
14. 4.63

WEIGHT

As explained earlier, there is some disagreement as to whether one should use the term *mass* or *weight* when referring to what one reads when stepping on the bathroom scales. This book recognizes the difference between these two terms but for everyday use recommends the common term *weight*. In technical and scientific work a clear distinction should be made.

Weight activities are placed after length activities in this book because weight materials such as scales and known weights are more difficult to obtain than metersticks and tape measures. Since the successful completion of a lesson for students without a feel for metric measures relies heavily upon the use of measurement devices, length activities should precede weight activities. Teachers inform the author repeatedly that they have little access to metric scales. This situation is slowly reversing itself. Among the few places where metric scales can be found are those schools where science programs which include weight apparatus have been purchased. Be sure and check around your building for metric scales that may be stored in unused science kits.

Numerous balance scales are available for purchase. Each classroom should have access to at least one commercially made scale (in the twenty-five dollar range) and weights. One scale is not enough, however, if the goal is to provide each student with weighing opportunities. There are various plans for making balance scales from milk cartons, coat hangers, etc. Such scales are satisfactory.

Hundreds of the following balance scales have been made and proved to be highly adequate. The cost of materials is approximately forty-five cents. The scale is sensitive and will easily weigh an object as light as 1 g. The only weakness is that this scale will weigh only items that fit into the cups. Some teachers have asked the local woodworking teacher to do the woodwork involved. Such a task might be welcomed by a student in need of a project.

Large arm balance

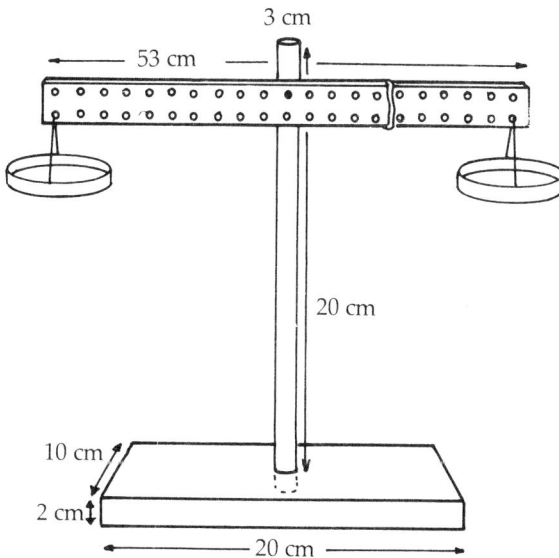

"*Materials:* Balance arm is made from pegboard and pivots on a nail through the stand. Cups are of Styrofoam, approximately 4 cm deep. Stand has a large dowel rod approximately 3 cm in diameter. Base is any suitable piece of wood. Rubber band may be moved in or out to balance arm."*

*From Kurtz, V. Ray: *Teaching Metric Awareness* (St. Louis: The C. V. Mosby Co., 1976).

Commercially made metric weights suitable for use in weighing are expensive and difficult to keep. These weights are usually made of brass and are quite attractive. Teachers report that they either get lost or walk off. An activity is presented in this section describing a procedure for preparing known weights. They are fun to make and work well in weighing activities.

Weight Diagnosis

In general, your students have had little experience with metric weights. You will probably be safe in assuming that they need beginning experiences with these units. You can easily diagnose their knowledge of metric weights, however, by giving the test that follows. You may wish to use the data from these tests in grouping class members for greater learning. Close answers should be considered correct.

MATERIALS: The test may be duplicated for distribution or it may be placed on the chalkboard.

NUMBER OF PARTICIPANTS: Entire Class.

DIRECTIONS: Estimate the metric weights of the following items. Have your teacher grade your answers.

Item	Estimated Weight
1. A nickel	
2. A regular M&M Candy	
3. A paper clip	
4. Yourself	
5. Your mathematics book	

Weight—M&M Weigh-in

Children love this activity. It may have something to do with the fact that they get to eat M&M candies. This activity is based on the fact that an M&M plain chocolate candy weighs very close to a gram. There is only one drawback—candies are expensive. Perhaps you have some source of funds for the purchase of edible materials.

MATERIALS: Balance scales, known 3 g weights and M&M plain chocolate candies.

NUMBER OF PARTICIPANTS: Any number of students working in pairs.

DIRECTIONS: "Each student may eat 3 grams of M&M candies after weighing them on the balance scales. How much does an M&M weigh?"*

Weight—Graham Cracker snack

This activity is designed to provide an edible experience for both younger and older learners. Since one square of graham cracker weighs just over 6 g, an opportunity is provided for the weigher to nibble a little of the cracker until it balances the 6 g weight.

MATERIALS: Balance scales, known gram weights and graham crackers.

NUMBER OF PARTICIPANTS: Any number.

DIRECTIONS: "Each student is to eat exactly 6 grams of graham cracker. This may be accomplished by snipping or breaking from one cracker until the amount remaining balances" a 6 g weight.**

Weight

Even though as adults we are bound to the foot/pound system, with our help students face no such mental blinders. By estimating and measuring they can be led to function as if the metric system were the only measurement system in existence. The following activity is designed to encourage estimation of weight. Whenever possible, estimations of the object should be made prior to the actual weighing. There should be a noted improvement between estimated weights and measured weights as the teams move through several stations.

MATERIALS: Metric scales and objects to weigh.

NUMBER OF PARTICIPANTS: Teams of three or four depending on the number of weighing stations.

*From Kurtz, V. Ray: *Teaching Metric Awareness* (St. Louis: The C. V. Mosby Co., 1976).

**From Kurtz, V. Ray: *Teaching Metric Awareness* (St. Louis: The C. V. Mosby Co., 1976).

DIRECTIONS: Place one object of unknown weight at each of several numbered weighing stations. Have each team move from station to station, first recording their estimated weight before weighing and recording the actual weight.

Weight

OBJECTIVE: To match decimal equivalency with metric prefix.

MATERIALS: Candy box and puzzle pieces prepared as shown in accompanying diagrams.

NUMBER OF PARTICIPANTS: Individual students.

DIRECTIONS: In the lid of a candy box, place the six metric prefixes. The decimal puzzle pieces are matched with the equivalent prefix and placed picture side up so that a picture will be formed when all pieces are correctly in place. This is a challenge to make but very rewarding for children to use.

milli	deka	kilo
deci	centi	hecto

Candy box with puzzle printed in the lid.

0.001	10	1,000
0.1	0.01	100

Bottom
Back side of picture when puzzle pieces are in place.

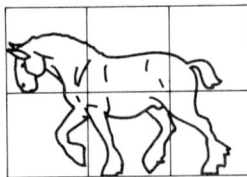

Picture when puzzle pieces are properly in place.

Weight

OBJECTIVE: To recognize metric weight terms.

MATERIALS: Copies of the metric maze.

NUMBER OF PARTICIPANTS: Any number.

DIRECTIONS: All the words listed can be found in the maze of letters. The words read forward, backward, up, down, and diagonally. Some words overlap and some letters are used more than once. Circle each word you find in the diagram and cross it off your word list.

metric	kilogram	gram	milligram
mass	hectogram	decigram	scales
weight	dekagram	centigram	

```
C E N T I G R A M V X K
R B N X F W E I G H T R
B D Z Y J D L W X X J L
O S M D Q L V B M G H Z
J L K A I F X A E G E S
M A R G R Z R Y T J C B
L G R S O G K C R B T O
T A G V A Z I D I M O J
M S R K P R N C C F G L
A T E N S C A L E S R V
S D Q B Z Y J D N D A G
S X R K I L O G R A M R
```

Weight—Metric Units

You should explain to your students that grams are used to measure the weight of small objects and kilograms are used to measure the weight of larger objects. It takes about 2 raisins to weigh 1 gram and 5 medium apples weigh about 1 kilogram.

MATERIALS: Copies of this table.

NUMBER OF PARTICIPANTS: Any number.

DIRECTIONS: What unit of metric weight would you use to measure:

Unit

1. A nickel		
2. A dollar bill		
3. A man		
4. A car		
5. A regular M&M candy		
6. Yourself		
7. Flour in a cake recipe		
8. A sack of candy		
9. A package of Kool Aid		
10. A box of Cracker Jacks		

ENRICHMENT ACTIVITY: Find the metric weight of the items by any method you may choose and record the weight in the last column.

Weight

OBJECTIVE: Stress extensive use of metrics in the everyday world.

MATERIALS: Copies of list.

NUMBER OF PARTICIPANTS: Any number.

DIRECTIONS: Take the following list to the grocery store and record the net weight in metric terms of each item.

1. Smallest can of pork and beans
2. Bag of marshmallows
3. Smallest jar of olives
4. Largest package of spaghetti
5. Largest bag of dried beans
6. Small box of Total breakfast cereal
7. Can of Pet milk
8. Largest loaf of bread
9. Smallest jar of jelly
10. Milky Way candy bar

Weight

OBJECTIVE: To weigh 1 gram amounts.

MATERIALS: Balance scales, borrowed items such as packing pieces, paper clips, sheet of paper, etc.

NUMBER OF PARTICIPANTS: Any number, depending on availability of equipment.

DIRECTIONS: See how many items you can find that weigh about 1 gram. Make a list of things that weigh 1 gram. You may need more than one of an item to weigh 1 gram.

Weight

OBJECTIVE: To arrange boxes in order according to weight.

MATERIALS: Four numbered boxes of similar size which contain 500 g, 1 kg, 1.5 kg, and 2 kg.

NUMBER OF PARTICIPANTS: Individual or small groups.

DIRECTIONS: Place the boxes in order from lightest to heaviest. Then estimate the weight of each box in grams or kilograms. Weigh each box to determine the closeness of your estimate.

Weight—Preparation of permanent weights

Permanent weights are expensive and difficult to keep. The following activity is used every year by one eighth grade teacher to insure that each student has a set of permanent weights for use throughout the school year. He reports that he has the learners permanently mark their initials on the weights so that when one is found on the floor, the owner can be identified.

MATERIALS: Balance scales, tin cans, tin snips or heavy scissors, and accurate weights of 1, 2, and 5 grams.

NUMBER OF PARTICIPANTS: Any number of students working in pairs.

DIRECTIONS: "Each team should cut from a tin can an amount of metal a little heavier than the known weight. Small amounts of material should be snipped from the tin until the scales indicate that the amounts are equal. In order to weigh up to 10 grams to the nearest gram, the set of weights needed would be one of 1 gram, two of 2 grams, and one of 5 grams. The numerical amount may be scratched on the metal piece, or if number punches are available, the number of grams may be punched into the weight. During the process of snipping, the sharp corner of the weights should be trimmed. The weights are not considered dangerous if proper precautions are given as to the sharpness of the metal edges. Children enjoy the experience of making the permanent weights. Vinyl floor tile may be used with younger children if the danger of being cut on the metal is considered to be too great."*

*From Kurtz, V. Ray: *Teaching Metric Awareness* (St. Louis: The C. V. Mosby Co., 1976).

Weight

The following activities are designed to be worked with the assistance of a hand-held calculator. The purpose of the activities is to demonstrate what happens when multiplying by the multiples and sub-multiples of 10.

MATERIALS: Hand-held calculators.

NUMBER OF PARTICIPANTS: Any number.

DIRECTIONS: Watch the display carefully after performing each of the following operations:

     1. Enter 3 g

        Multiply by 10
        Multiply by 10
        Multiply by 0.1
        Multiply by 0.1
        Multiply by 100
        Multiply by .01

        What do you have remaining? (3 g)

     2. Enter 3.333 g

        Multiply by 10
        Multiply by 10
        Multiply by 10
        Multiply by 10
        Multiply by 0.1

        What do you have remaining? (3333 g)

Weight—Gram-o-rummy

This card game is designed to provide reinforcement of the metric units of weight. Each of the seven units—kg, hg, dag, g, dg, cg, and mg—is used in the game. All need to be included in order to stress that the pattern of the prefixes is similar to the pattern of place value and money.

MATERIALS: Twenty-eight cards prepared with the following designations: four with mg, four with dg, four with g, four with dag, four with hg, four with kg, and four with cg.

NUMBER OF PARTICIPANTS: Two or three (more may play if additional cards of each unit are prepared).

DIRECTIONS: "Deal 5 cards to each player. Place the remaining cards face down in a reserve stack and turn the top card over. This becomes the discard stack. The player to the left of the dealer begins by either taking the top card in the discard stack or drawing a card from the reserve stack. The object of the game is to make a run of 5 cards in consecutive order beginning at any point. Each player in turn, moving counterclockwise, either draws a card from the reserve stack or takes the top card from the discard stack. He discards a card from his hand each time he plays. The first player to get a run, such as kg, hg, dag, g, dg; or dag, g, dg, cg, mg, lays his cards down and he is declared the winner. (Children won't stop playing this game.)"*

Weight

OBJECTIVE: To state metric weight terms in a puzzle.

MATERIALS: Copies of the accompanying puzzle.

NUMBER OF PARTICIPANTS: Any number.

DIRECTIONS: Complete the puzzle.

Across

1. A metric _____ is 1000 kg.
4. A nickel weighs approximately _____ g.
5. The tar in cigarettes is measured in _____.
7. Six milliliters of water would weigh _____ grams.
8. My mother weighs _____ kilograms.

Down

2. A dollar bill weighs _____ gram.
3. Heavy objects are measured in _____.
5. Sometimes we use the term _____ instead of weight.
6. The symbol for milligram is _____.

*From Kurtz, V. Ray: *Teaching Metric Awareness* (St. Louis: The C. V. Mosby Co., 1976).

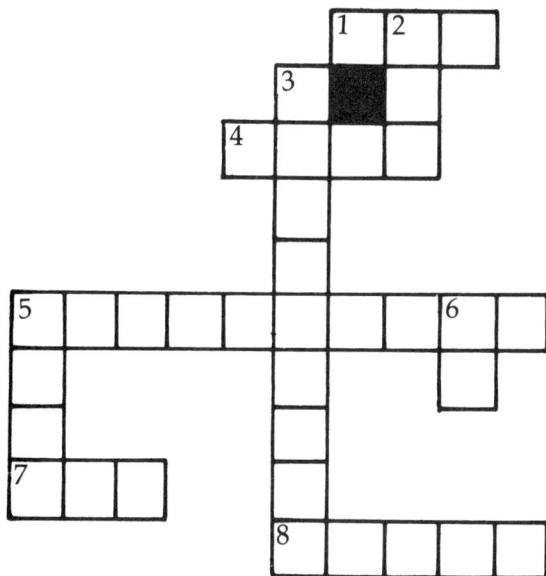

Answers:

Across

1. ton
4. five
5. milligrams
7. six
8. sixty

Down

2. one
3. kilograms
5. mass
6. mg

Weight

Abstract activities such as the following should follow after numerous concrete activities have been completed. After the student has gained a feel for the gram, milligram, and kilogram, he or she is ready to practice changing from one weight unit to another.

You will find that showing students the following chart will aid in the conversions.

kg	hg	dag	g	dg	cg	mg

In looking at the chart you see that when moving from grams to kilograms, you move three positions (or decimal places) to the left. Therefore, in the first problem the 1000 g would change to 1 kg. Just reverse the process when moving from a larger unit to a smaller, e.g., 2 kg would be 2000 g.

MATERIALS: Copies of the following problems.

NUMBER OF PARTICIPANTS: Any number.

DIRECTIONS: Change units as directed.

1. 1000 g   = _____ kg
2. 1000 mg   = _____ g
3. 1342 mg   = _____ g
4. 1.34 kg   = _____ g
5. 0.7 kg   = _____ g
6. 17 kg   = _____ g
7. 34 000 mg   = _____ g
8. 7 kg   = _____ g
9. 3400 g   = _____ kg
10. 29 g   = _____ mg

11. 3.9 g   = _____ mg
12. 4.23 g   = _____ mg
13. 200.7 mg   = _____ g
14. 70.63 mg   = _____ g
15. 239.6 g   = _____ mg
16. 3422.96 g   = _____ kg
17. 272 g   = _____ mg
18. 723 mg   = _____ g
19. 0.027 kg   = _____ g
20. 0.001 kg   = _____ g

Answers:

1. 1 kg
2. 1 g
3. 1.342 g
4. 1340 g
5. 700 g
6. 17 000 g
7. 34 g
8. 7000 g
9. 3.4 kg
10. 29 000 mg

11. 3900 mg
12. 4230 mg
13. 0.2007 g
14. 0.070 63 g
15. 239 600 mg
16. 3.422 96 kg
17. 272 000 mg
18. 0.723 g
19. 27 g
20. 1 g

AREA

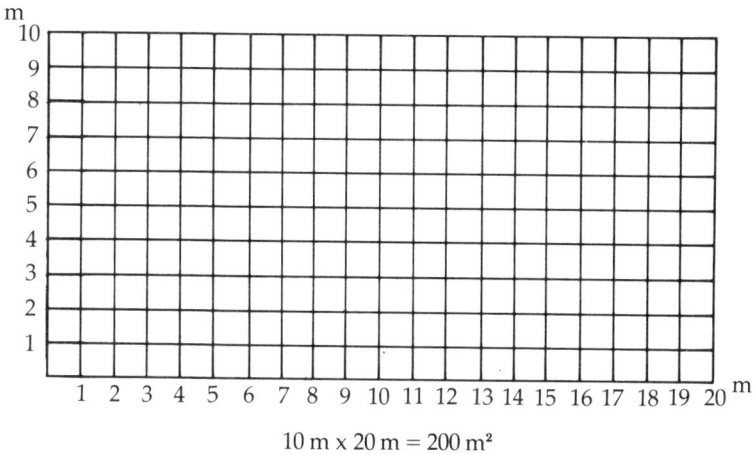

$$10 \text{ m} \times 20 \text{ m} = 200 \text{ m}^2$$

It is not uncommon for adults to be confused between square units and cubic units. This confusion exists for two reasons. First, many times measurement activities are not stressed as much as computation-type problems. There is reason to believe that geometry pages are among the most frequently skipped pages in mathematics books. Teachers sometimes feel that there just isn't time for everything and therefore something must go. Many times measurement goes.

A second reason for the confusion between square and cubic units is that we rely almost completely on the printed page to teach area and volume. The two-dimensional page does not adapt well to three-dimensional shapes. This confusion becomes quite obvious when a learner says, "Now let's see, do you measure the area of this surface in square centimeters or cubic centimeters?" Anyone, young or old, making such a statement needs concrete experiences with area and volume. Cuisenaire rods or similar rods are an excellent place to start.

The most commonly used metric units of area are as follows:

Unit	Symbol
Square centimeter	cm²
Square meter	m²
Are (pronounced *air*)	a
Hectare	ha

The *hectare* is a unit of land measurement equal to 100 *ares*. The *are* is equal to a square *dekameter*. To demonstrate the size of an are, a square ten large steps on each side may be paced off. A hectare would be equal to 100 ares. This will give learners an idea of the large size of a hectare.

Area

OBJECTIVE: To gain a feel for the size of a square centimeter.

MATERIALS: Cuisenaire rods or other metric-sized cubes and rods.

NUMBER OF PARTICIPANTS: Any number.

DIRECTIONS: Take a white rod from your pile and press one side against your palm. Lift the rod and you will find an imprint of a square centimeter. How many square centimeters are showing (surface area) on a white rod? (6 cm²)

What is the surface area of a red rod? (10 cm²)

What is the surface area of a light green rod? (14 cm²)

What is the surface area of two yellow rods placed long sides together? (34 cm²)

Area

During the past decade, geoboards have become very popular for reducing difficult abstract geometric concepts to the concrete level. The area and perimeter of a geometric shape can be determined without the use of a formula. Concrete geoboard activities are excellent for developmental activities in introducing the concepts of area and volume. The biggest problem is that virtually none of the geoboards in classrooms is metric. The geoboard described in this activity is readily constructed from the materials listed.

MATERIALS: Ceiling tile, common pins, and string.

NUMBER OF PARTICIPANTS: Any number.

DIRECTIONS: Prepare a metric geoboard in the following manner. Draw a 1 cm grid on the finished side of a ceiling tile or lay a sheet of centimeter grid paper on the tile. Place a common pin at each line intersection. When finished, you will have a pin in each corner of each square centimeter. After finishing the geoboard, use string or a rubber band to place around the pins making a triangle, a square, a rectangle, and a trapezoid.

Area

The first experiences for learners in fourth and fifth grades should involve using grid paper or a grid transparency. The use of formulas should not be used initially. This activity is designed to use a grid transparency in the determination of the area of an object. The transparency may be made by running a sheet of 1 cm grid paper and a blank transparency through a Thermofax machine. The objects to be measured should be rectangles or squares that contain a whole number of square centimeters. Some students may discover that the area of a square or rectangle may be determined without counting all the square centimeters. This activity may be used to lead directly into the use of the area formula.

MATERIALS: One centimeter grid transparency and one sheet of square and one of rectangular shape.

NUMBER OF PARTICIPANTS: Any number.

DIRECTIONS: Place the grid over the shape to be measured. Count the square centimeters contained in the area of the shape. Check your answer with a neighbor. If you do not agree, count the square centimeters again.

Area

This area activity is designed to follow the preceding exercise. This time the learner will be measuring shapes that do not fit evenly under the grid. A sheet of right triangles, circles and uneven shapes will be measured. The student must estimate the partial square centimeters when counting the total. He or she may discover that the area of a right triangle is one-half the area of a square or rectangle of the same dimensions.

MATERIALS: Centimeter grid transparencies and sheets of right triangles, circles, and uneven shapes.

NUMBER OF PARTICIPANTS: Any number.

DIRECTIONS: Place the grid on the shape to be measured. First count the number of whole square centimeters under the grid. Then estimate the partial square centimeters under the grid. Add these two numbers together. This should approximate the area of the shape. Check your answer with a friend.

Area

OBJECTIVE: To measure foot area.

MATERIALS: Grid paper.

NUMBER OF PARTICIPANTS: Any number.

DIRECTIONS: Using centimeter grid paper, trace around your feet with shoes removed. Count the number of square centimeters within the boundary of the outline of your feet. First add the whole squares, then estimate the partial squares. After adding together the area of each foot, check with your classmates to see who has the greatest foot area.

ENRICHMENT ACTIVITY: Determine the weight on each square centimeter of your two feet. Again check with your classmates to see who has the most pressure in kilograms or grams per square centimeter. Ask the person who has the most pressure if he or she has sore feet.

Area

OBJECTIVE: To measure the area of your hand in square centimeters.

MATERIALS: Grid paper.

NUMBER OF PARTICIPANTS: Any number.

DIRECTIONS: Using centimeter grid paper, trace around your hand placed palm down with fingers and thumb together. Count the number of square centimeters encompassed in the outline of the hand. Stop counting at the line where the wrist bends. Which has the greatest area—your foot, as measured in the preceding activity, or your hand? After checking with friends, could you predict the area of one if you knew the area of the other?

Area

The geoboard described earlier (made with ceiling tile and common pins) is easily constructed but not as durable as the one described here. This geoboard is constructed from scrap wood and nails. A piece of wood at least 10 cm square will make an excellent geoboard and should be readily available from the scrap bin at lumber yards. Four penny finish nails are spaced 2 cm apart. The grid should be drawn in with a ball point pen because of the 2 cm spacing of the nails. After preparing a few of these geoboards, try the following activities.

MATERIALS: Geoboards as described.

NUMBER OF PARTICIPANTS: Any number depending on geoboards available.

DIRECTIONS: 1. Construct a triangle of 2 cm^2, 4 cm^2, 4 cm^2 and 6 cm^2
2. Construct a square with an area of 4 cm^2
3. Construct a rectangle with an area of 8 cm^2
4. What is the area of a right triangle with a base of 2 cm and height of 4 cm? What happens to the area if you double the base to 4 cm and keep the height at 4 cm? What happens if you double both the base and height?
5. Construct a square that contains 8 cm^2. (It can be done.)

Area

OBJECTIVE: To gain a feel for the size of an *are* and a *hectare*.

MATERIALS: Meterstick or tape measure.

NUMBER OF PARTICIPANTS: A team of four members.

DIRECTIONS: Measure a square that is 10 meters or 1 dekameter on each side. You may do this in your classroom if it is large enough. If it is not large enough, measure the square in the cafeteria, gymnasium or school ground. Imagine the size of 100 ares or a hectare.

ENRICHMENT ACTIVITY: Measure the school ground in hectares. Challenge another team or
another class to measure the
school ground. Compare your
answer with theirs.

Area (Advanced)

OBJECTIVE: To state the area of various-shaped triangles
when the base and height remain constant.

MATERIALS: Geoboards or grid paper.

NUMBER OF PARTICIPANTS: Any number.

DIRECTIONS: 1. Use your geoboard and make a right triangle
with a base of 4 cm and height of 2 cm. What
is the area of this triangle? (Answer: 4 cm²)

2. Again use your geoboard and construct an
isosceles triangle with a base of 4 cm and a
height of 2 cm. What is the area of this
triangle? (Answer: 4 cm²)

3. Construct another triangle of different
shape with the same base and the same
height. What is the area of this triangle?
(Answer: 4 cm²) What generalization can you
make about the area of triangles of different
shapes if the base and height remain the same
values? (Answer: Area will remain the same.)

Area (Advanced)

The following activity may be worked using the
hand-held calculator. The problems are not difficult,
but the calculator will provide some added incentive to
working them. If you do not have enough calculators
for total class participation, you may want to have a
learning center or interest corner where these and
similar problems can be worked by students on an
independent basis.

MATERIALS: Hand-held calculators.

NUMBER OF PARTICIPANTS: Depends on availability of hand-held calculators.

DIRECTIONS: Using a hand-held calculator:

1. Find the surface area of a rectangular solid that is 12 cm by 24 cm by 29 cm. (Answer: 2664 cm^2)

2. Find the surface area of a rectangular solid that is 15 cm by 30 cm by 48 cm. (Answer: 5220 cm^2)

Area (Advanced)

OBJECTIVE: To state the area of various-shaped rectangles or squares when the perimeter remains the same.

MATERIALS: Geoboards or grid paper.

NUMBER OF PARTICIPANTS: Any number.

DIRECTIONS: If you were given 36 meters of fencing to enclose a garden in the shape of a rectangle or a square, what shape would you make the garden so as to have the greatest area? Does the area vary as the shape of the garden varies? What shape would you make the garden if you wanted the least area? Remember, you are to use all the fencing in making the garden. No dimension should be less than 1 meter. (Answer: Garden of largest area would be a square 9 m by 9 m. The garden of least area would be 1 m by 17 m.)

Area (Advanced)

OBJECTIVE: To compute the number of hectares when dimensions are given.

MATERIALS: Following table.

NUMBER OF PARTICIPANTS: Any number.

DIRECTIONS: Compute the number of hectares in each field in the following table:

Field	Dimensions	Area in Hectares
1	100 m x 100 m	
2	50 m x 500 m	
3	20 m x 80 m	
4	1 km x 500 m	
5	1.3 km x 300 m	

Answers:

1. 1 ha
2. 2.5 ha
3. 0.16 ha
4. 50 ha
5. 39 ha

VOLUME—CAPACITY

In the middle 1700s English schoolchildren were taught to memorize the following units:

Two mouthfuls are a jigger; two jiggers are a jack; two jacks are a gill; two gills are a cup; two cups are a pint; two pints are a quart; two quarts are a pottle; two pottles are a gallon, two gallons are a pail; two pails are a peck; two pecks are a bushel; two bushels are a strike; two strikes are a coomb; two coombs are a cask; two casks are a barrel; two barrels are a hogshead; two hogsheads are a pipe; two pipes are a tun—and there my story is done!

How would you like that system of measurement?

Volume—Capacity

The difference between the terms *volume* and *capacity* is slight. Some educators advocate not distinguishing between the terms. Even though it may not be necessary for young learners to distinguish, it is necessary that teachers know the difference.
The following container shows the difference:

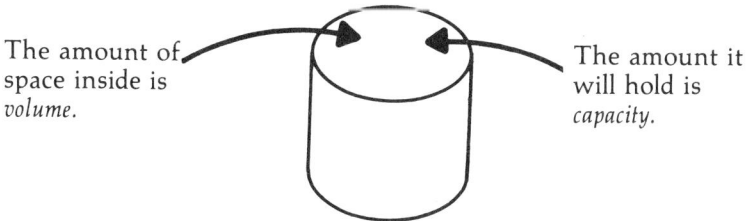

The amount of space inside is *volume*.

The amount it will hold is *capacity*.

Cubic centimeters usually indicate the volume of solids or the size of a product. *Liters* and *milliliters* are normally used when measuring liquids or the capacity of a container to hold liquids. A milk carton may have a volume of 1000 cm³ but a capacity to hold 1 L of milk.
As stated earlier, both young and old learners have difficulty conceptualizing the relationships between volume, capacity, and weight. These relationships are shown in the following table:

Volume	Capacity	Weight (water only)
1 cm³	1 mL	1 g
1 dm³	1 L	1000 g
1 m³	1 kL	1 metric ton

From the table you can see that a container with a volume of 1 cm³ has a capacity of 1 mL of water. This amount of water will weigh 1 g. The three volumes listed, *cm³*, *dm³*, and *m³*, should be used in readiness activities for volume and capacity. The white Cuisenaire rod may be used to model a cubic centimeter. The paper centimeter cube described as an activity in this section may be constructed. Cubic decimeters are readily available from commercial supply houses or can be made from one-half gallon milk cartons as described in a later activity. Constructing the cubic meter presents a somewhat more difficult situation. It is recommended that each school purchase a cubic meter for classroom use. If this is impossible, twelve metersticks can be taped together to form a cubic meter. Since all international trading of grain is done in metric tons, relating the cubic meter to the metric ton of water is a highly desirable activity.

Even though capacity is defined by volume, the concept of capacity should be taught earlier than the concept of volume. You may use activities that stress the basic unit of capacity, the liter, as early as second grade. You should make sure that your learners understand the concept of capacity, however, before teaching the liter. If they do not understand this concept, you should first use activities involving nonstandard measurement many of which are provided in this book.

Capacity

Young learners who are just conceptualizing the capacity of a milliliter and a liter need to experience these measures in situations that are common in their lives. This activity is designed to provide that experience. Some soft drink companies were early in marketing a liter bottle. This bottle provides the vehicle to use in this activity. Since many students have not equated the number of milliliters in a liter, ask them to determine this quantity. Any graduated container marked in milliliters may be used. This activity may be followed by measuring the milliliters of soda in normal-sized containers dispensed by coin-operated machines.

MATERIALS: Several different soft drink bottles including a liter bottle and a graduated cylinder.

NUMBER OF PARTICIPANTS: Teams of three or four members.

DIRECTIONS: Fill a one-liter soda bottle with water. Measure in milliliters the capacity of the bottle by pouring the water into a graduated container. After completing this activity, calculate the milliliters in common-sized soda containers. Ask learners how many milliliters of soda or water they can drink.

Capacity

The day before you plan to make liter containers you will want to ask each student to bring a one-half gallon milk carton to school the next day. If the container is prepared in the following manner, it will hold very close to 1 liter of water or whatever liquid you may want to put in it.

MATERIALS: One-half gallon milk cartons, X-acto knife or scissors, and a centimeter ruler.

NUMBER OF PARTICIPANTS: Any number.

DIRECTIONS: Measure 10.5 cm from the bottom of the milk container and mark a line around the carton. Carefully follow the line and cut the carton until the top portion of the container is free. The bottom part of the container will measure approximately 9.5 cm wide, 9.5 cm deep and 10.5 cm high. Even though this is not precisely the 10 cm^3 which by definition will hold a liter, it is very close. Ask each student to 1) measure his or her liter container and 2) fill the container with water and pour it into a friend's container to see if they hold the same amount.

Capacity

Your students must "feel" metric before they can "think" metric. The metric language used in class should constantly be applicable to real situations existing in the lives of the learners. The *milliliter* and *liter* are the units most often used in measuring capacity. Soft drinks and medicines are measured in milliliters, while oil and gasoline are measured in liters.

MATERIALS: Exercise as listed.

NUMBER OF PARTICIPANTS: Any number.

DIRECTIONS: Which unit of capacity, the milliliter or liter, would you use to measure the following:

1. Can of juice
2. Gasoline tank of car
3. Fishbowl
4. Bathtub
5. Cough syrup
6. Milk
7. Water jug
8. Coffeepot

Capacity

This activity is designed for use in helping children gain an understanding of *capacity*. The element of team competition is included to provide added interest. The winning team should receive a certificate listing their names which should be placed in the classroom display area.

MATERIALS: Ten containers such as coffee cans, fruit cans, juice cans, jars, and bottles.

NUMBER OF PARTICIPANTS: Participating teams of five members each and one team of three judges.

DIRECTIONS: Identify each container with a number. Ask each team to arrange the containers in order from smallest to largest by estimation. Instruct the group of judges to carefully measure each container to determine the capacity in milliliters. The winning team will be designated as such. A losing team may challenge a measurement and check it themselves.

Capacity

OBJECTIVE: To measure in milliliters the amount of milk in a school-sized milk carton.

MATERIALS: Milk cartons, graduated cylinder and water.

NUMBER OF PARTICIPANTS: Any number.

DIRECTIONS: Measure the milliliters of milk that you drank for lunch. Do so by filling an empty carton with water to the same level that it was filled with milk. Then pour the water into a graduated cylinder to determine the number of milliliters in the milk carton. If the carton states the capacity in milliliters, verify this measurement with your experiment.

Capacity

Even though car owners commonly drive up to gas stations and say, "Fill it up," or "Give me 10 gallons," many times they are unaware of the actual amount of gas purchased because the tank is hidden under the fender or trunk. The following activity is designed to prepare the student for the day when gasoline is purchased by the liter. To carry it out a rather large container such as a tub, barrel, etc., is needed. Two or more smaller containers may be used.

MATERIALS: One or more containers that will hold 50 liters of water and a liter container.

NUMBER OF PARTICIPANTS: A team of four or five students.

DIRECTIONS: Measure 50 liters of water and pour it in one or more containers. Could this correspond to the amount of gasoline purchased for your family automobile?

Capacity

Abstract activities such as the following should be given after completing numerous concrete activities. When the student has gained a feel for the *liter* and *milliliter,* he or she should practice changing from one capacity unit to another.

MATERIALS: Copies of the following problems.

NUMBER OF PARTICIPANTS: Any number.

DIRECTIONS: Change units as directed.

1. 3 L = _____ mL	8. 0.007 L = _____ mL
2. 3.4 L = _____ mL	9. 0.146 L = _____ mL
3. 0.7 L = _____ mL	10. 1.342 L = _____ mL
4. 0.14 L = _____ mL	11. 0.93 mL= _____ L
5. 300 mL = _____ L	12. 0.72 mL= _____ L
6. 3042 mL= _____ L	13. 0.723 L = _____ mL
7. 34.96 mL= _____ L	14. 300 L = _____ mL

Answers:

1. 3000 mL	8. 7 mL
2. 3400 mL	9. 146 mL
3. 700 mL	10. 1342 mL
4. 140 mL	11. 0.000 93 L
5. 0.3 L	12. 0.000 72 L
6. 3.042 L	13. 723 mL
7. 0.034 96 L	14. 300 000 mL

Capacity—Blow a liter of hot air

This is an advanced activity for older learners. If the directions are given clearly, the activity can be accomplished by fifth graders and older students. As teacher, you will want to become thoroughly familiar with the activity before presenting it to the class. As you give directions, you should move through the class so that difficulties can be spotted and help can be given. Only newsprint will work well in this activity.

MATERIALS: Sheets of newsprint that form a square 40 cm on
               each side.

NUMBER OF PARTICIPANTS: Entire class.

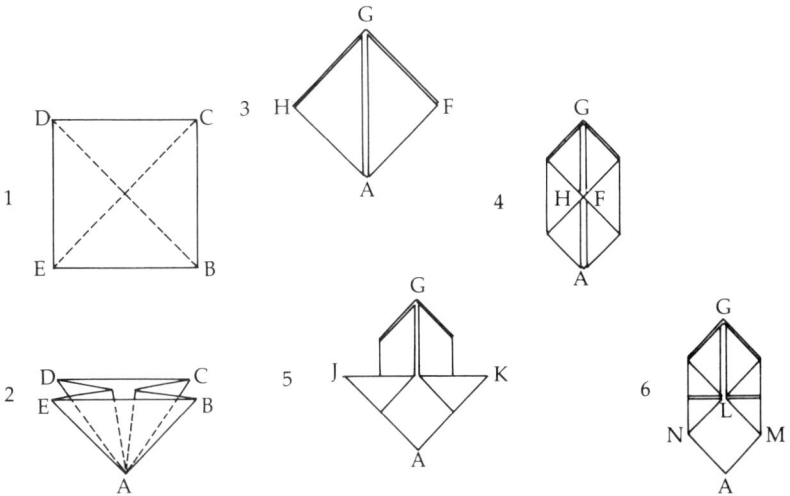

DIRECTIONS: "1. Fold the paper from corner to corner as in Diagram 1.
2. Fold the sides in to form a triangle as shown in Diagram 2.
3. Fold points E and B down to point A. Turn it over and fold C and D down to point A. Keep point A pointing downward as in Diagram 3.
4. Corners H and F are now double and A is loose. Fold H and F to meet in the center. Turn it over and do the same for the corners on the back, Diagram 4.
5. Fold the loose ends out to form Diagram 5. Turn it over and do the same on the back.
6. Fold J and K over to the middle to form right triangles. Turn it over and do the same for the back, Diagram 6.
7. Tuck the small right triangles made in number 6 into the pockets along LM and LN. Do not tuck them under LM and LN. Turn the model over and do the same on the back.
8. Open the hole at G slightly with a pencil point. Blow hard into this hole, forming and creasing the cube as you blow. The finished cube will be 10 centimeters on a side and will hold 1 liter of air."*

*From Kurtz, V. Ray: *Teaching Metric Awareness* (St. Louis: The C. V. Mosby Co., 1976).

Volume-Capacity—Word Hunt

OBJECTIVE: To identify metric volume words.

MATERIALS: Newspapers,   magazines,   food   boxes   and
                     wrappers.

NUMBER OF PARTICIPANTS: Any number.

DIRECTIONS: Challenge the class to a word hunt by telling
                 them that the person (or team) to find the most
                 metric words expressing volume or capacity will
                 be awarded a certificate. The hunt will be for the
                 following words and symbols: liter, milliliter,
                 cm^3, and m^3. The logical places to look for these
                 words would be in newspapers, magazines, and
                 on food containers and wrappers. No foods—
                 only the empty containers—may be brought to
                 school.

Volume-Capacity—Scavenger Hunt

OBJECTIVE: To provide experiences selecting metric volume
                 and capacity.

MATERIALS: Various items as suggested.

NUMBER OF PARTICIPANTS: Any number.

DIRECTIONS: Go and find the following items:

           1. Something that holds more than a liter of
              liquid.

           2. Something that contains a volume of less
              than 20 cm^3.

           3. Something that holds exactly 2 liters.

           4. Exactly 1 mL of water.

           5. A food container with contents given in
              milliliters.

Volume

OBJECTIVE: To construct a cubic centimeter.

MATERIALS: Rubber cement, paper and scissors.

NUMBER OF PARTICIPANTS: Any number.

DIRECTIONS: Copy the accompanying pattern. Be sure that each side is 1 cm in length. After cutting out the pattern, fold to form a cube. The tabs are folded so they can be glued to the outside of the adjoining sides. Place rubber cement on the tabs before putting them in place. The cube will have a capacity of 1 mL. If filled with water it would weigh 1 g. You may want to try the following related activities.

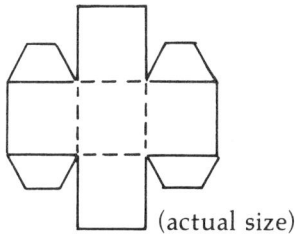

(actual size)

1. If your cube is made very carefully, it will hold water. You may need to dab a little rubber cement on the corners to make it watertight. Weigh the cube full of water to see if it weighs close to 1 g. Don't forget to subtract the weight of the paper cube.

2. Add another side to enclose the cube.

3. Make a cubic decimeter with an open top and ask your friends to help you fill it with cubic centimeters. How many will it hold?

Capacity

OBJECTIVE: To state metric terms of capacity in a puzzle.

MATERIALS: Copies of the accompanying puzzle.

NUMBER OF PARTICIPANTS: Any number.

DIRECTIONS: Complete the puzzle.

Across

1. There are one _____ mL in a L.
3. The basic units of capacity are _____.
6. A _____ is .001 of a liter.

Down

2. Medicines and _____ are bought by the milliliter.
3. _____ are measured by capacity.
4. The symbol for milliliter is _____.
5. We will buy cartons of _____ by the liter.

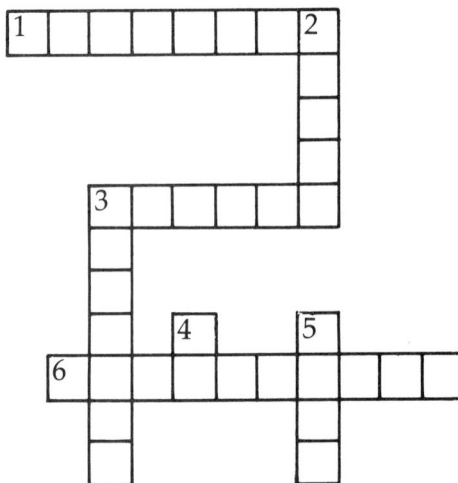

Answers:

Across

1. thousand
3. liters
6. milliliter

Down

2. drugs
3. liquids
4. mL
5. milk

Volume-Capacity—Weight (Advanced Activity)

The relationship between volume, capacity and weight is difficult for young and old alike to grasp. The following activity will present in a concrete way the concept that one cubic centimeter will hold one milliliter of water and will weigh one gram.

MATERIALS: Finely calibrated graduated cylinder, clay, paper cup, water, and balance scale.

NUMBER OF PARTICIPANTS: Teams of three students.

DIRECTIONS: Using tightly packed clay, form centimeter cubes. Drop these cubes in a graduated cylinder containing 50 mL of water. An example of the concept of displacement, these cubes will displace 1 mL of water when dropped into the cylinder. The rise in water level can be observed for each cube dropped into the water. After it is shown that 1 cm^3 of clay raises the level 1 mL, the final phase of weighing 5 mL or more of water may take place. Weigh a paper cup before adding the water. The weight in grams of the cup and water minus the weight of the cup will equal the number of milliliters of water displaced by the cubes of clay.

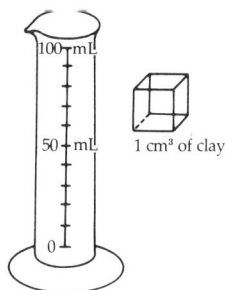

1 cm^3 of clay

Volume, Capacity, and Weight
Relationships of Metrics

You will recall that meters, liters, and grams are all related. Following are the important relationships.

    a. One cubic centimeter of water is equivalent to 1 milliliter of water and has a weight of 1 gram.

b. One cubic decimeter of water is equivalent to 1 liter of water and has a weight of 1 kilogram.

c. One cubic meter of water is equivalent to 1 kiloliter of water and has a weight of 1 metric ton.

OBJECTIVE: To demonstrate the relationships between metric terms of volume, capacity and weight.

MATERIALS: Copies of table.

NUMBER OF PARTICIPANTS: Any number.

DIRECTIONS: Use the facts given to complete the table.

Volume	1 cm^3	1 dm^3				2 m^3
Capacity	1 mL		5 L			
Weight (water)	1 g			75 g	6.5 g	

ENRICHMENT ACTIVITY: Bill has an aquarium that is 0.8 m x 0.3 m x 0.4 m

1. What is the volume of the aquarium in cubic centimeters?

2. What would be the weight of the water in kilograms if the aquarium were filled?

Answers:

Volume	1 cm^3	1 dm^3	1 dm^3	75 cm^3	6.5 cm^3	2 m^3
Capacity	1 mL	1 L	5 L	75 mL	6.5 mL	2 kL
Weight (water)	1 g	1000 g	5000 g	75 g	6.5 g	2 metric tons

1. 96 000 cm³
2. 96 kg

Volume (Advanced)

Students are encouraged to use the hand-held calculator in computing the volume of the following sphere. This problem provides experience in using the decimal equivalent of the fraction 4/3. Students should be reminded that the final answer will only be an approximation (even though a very close one).

MATERIALS: Hand-held calculator.

NUMBER OF PARTICIPANTS: Any number.

DIRECTIONS: Find the volume of a sphere that has a radius of 30 cm. The formula is: $V = 4/3 \pi r^3$. Use 3.1416 for π. (Answer: 113 097.59 cm³)

TEMPERATURE

	C°
WATER BOILS	100°
EXTREMELY HOT	40°
BODY TEMPERATURE	37°
BEACH WEATHER	25°
COMFORTABLE ROOM	22°
WATER FREEZES	0°
COLD	−6°
VERY COLD	−18°
BITTER COLD	−30°
	−40°
	−273°

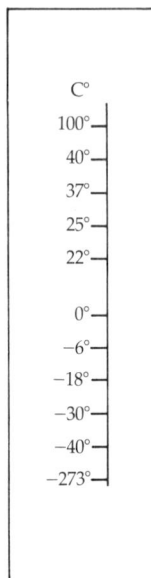

Even though the SI basic unit of temperature is the *kelvin degree*, the Celsius scale is used extensively in countries that use the metric system of measurement. The kelvin scale is oftentimes used in scientific work, while the Celsius scale is used in practical everyday situations. As mentioned previously, the centigrade scale is identical to Celsius. The latter is named after Anders Celsius who is reported to be the first person to use a thermometer with 100 degrees between freezing and boiling water. All three scales, kelvin, Celsius, and centigrade have identical units, i.e., a one degree change in temperature on one scale is equivalent to a one degree change on the other scale. The kelvin thermometer has an absolute zero, at which temperature, molecular and atomic motion stop.

The changeover from thinking in degrees Fahrenheit to degrees Celsius is not accomplished overnight. Learners of the new system need many experiences which require the use of Celsius thermometers. Textbook metrics used in the past will not give the student a feel for metric temperatures. The following broad guidelines will help in conceptualizing these temperatures:

tingling tens

temperate twenties

thirsty thirties

flaming forties

As you teach the Celsius scale, you will soon learn that negative temperatures occur more often than with the

Fahrenheit scale. On many winter days when the temperature dips to 30°F, 20°F, or 10°F, the corresponding Celsius temperatures would be negative readings. Consequently, young students will need earlier experiences with the set of integers than in the past.

The following activities are designed to develop in the learner the ability to "think Celsius."

Temperature Diagnosis

The following questions may be used as a pre-test or post-test to the temperature unit. Students should be able to answer four out of six questions correctly after completing a temperature unit.

MATERIALS: Questions as stated.

NUMBER OF PARTICIPANTS: Any number.

DIRECTIONS: Answer the following questions:

1. If the thermometer in your house reads 20°C in the winter, will you be comfortable?

2. If the thermometer reads 5°C, will you need a short-sleeved shirt or a jacket?

3. If the thermometer reads 25°C, would it be pleasant on the beach?

4. If your bathwater measures 20°C, will you have a hot, warm, or chilly bath?

5. If the temperature reaches 35°C tomorrow will it be cool, warm, or hot?

6. Will the snow melt if it is 10°C?

Answers:
1. yes   4. chilly
2. jacket   5. hot
3. yes   6. yes

Temperature

OBJECTIVE: To provide learners with the opportunity of reading a Celsius thermometer.

MATERIALS: Page of duplicated thermometers.

NUMBER OF PARTICIPANTS: Any number.

DIRECTIONS: Read the temperatures as recorded on the following thermometers:

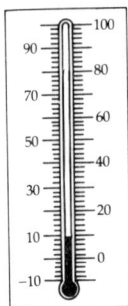

_____ °C

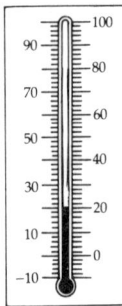

_____ °C

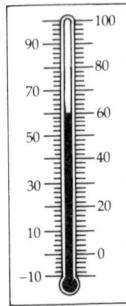

_____ °C

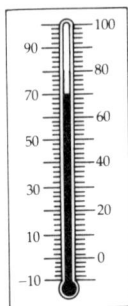

_____ °C

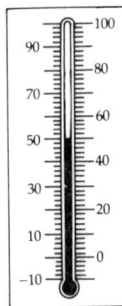

_____ °C

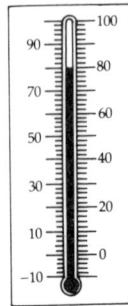

_____ °C

Temperature

A "feel" for metric temperatures in degrees Celsius is not attained easily. To gain this competency, students will need many encounters with temperature, both physical and in class discussions. The following temperature recording exercise is one of such preliminary activities that must be planned and carried out. The thermometer needed for this activity is readily available at most hardware stores. If it has both Fahrenheit and Celsius scales, the Fahrenheit scale may be blocked out with tape. School supply houses have inexpensive Celsius thermometers.

MATERIALS: Celsius thermometers and the accompanying chart.

NUMBER OF PARTICIPANTS: Teams of two.

DIRECTIONS: Use the following chart to record the temperature each day at 9:00 A.M. Another team should use a different copy of the chart to record the temperature at 12 noon. A third team should use an even different chart to record the temperature at 3 P.M.

Day	Outside Temperature in °C	Day	Outside Temperature in °C
Monday		Monday	
Tuesday		Tuesday	
Wednesday		Wednesday	
Thursday		Thursday	
Friday		Friday	

At the end of each week make a line graph of the temperature. Solid, colored, or dotted lines may be used.

Temperature

OBJECTIVE: To become familiar with metric temperatures.

MATERIALS: Magazines, scissors, glue and paper.

NUMBER OF PARTICIPANTS: Any number.

DIRECTIONS: After discussing in degrees Celsius the outside temperature, inside temperature, ice temperature, body temperature and water boiling temperature, children should be instructed to cut pictures from magazines and glue them on sheets of paper. Realistic temperature in degrees Celsius should be recorded under the pictures. Summer pictures will have temperatures of 25°C to 35°C. Winter pictures will have temperatures of minus 20°C to 5°C. Spring and fall temperatures will range from 10°C to 25°C.

Temperature

The following activity is designed to provide the learner with an opportunity to practice using temperature in degrees Celsius. This activity should take place independent of the Fahrenheit scale. The objective is to think in degrees Celsius, not degrees Fahrenheit. This cannot be accomplished by constantly referring to Fahrenheit readings and comparing them with Celsius.

MATERIALS: Accompanying thermometer and list of questions.

NUMBER OF PARTICIPANTS: Any number.

DIRECTIONS: Use the accompanying thermometer to answer the following questions:

_____ 1. How many degrees between the freezing and boiling points of water?

_____ 2. If your body temperature were 39°C, you would have how many degrees of fever?

_____ 3. The temperature was 50°C and dropped 35°. What was the resulting temperature?

_____ 4. The temperature was 13°C and then gained 15°. What was the resulting temperature?

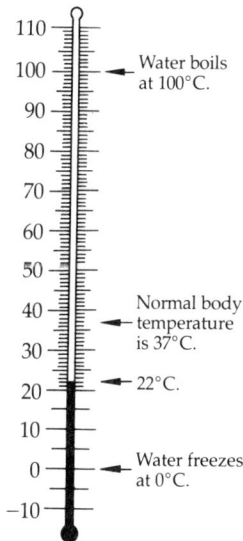

Temperature

OBJECTIVE: To state metric terms for temperature in a puzzle.

MATERIALS: Copies of the accompanying puzzle.

NUMBER OF PARTICIPANTS: Any number.

DIRECTIONS: Complete the puzzle.

Across

2. We measure temperature in _____.
3. Thirty-seven degrees Celsius is _____ temperature.
6. The _____ at which snow melts is 0°C.
7. The _____ point of water is 0°C.

Down

1. We measure temperature with a _____.
4. The _____ point of water is 100°C.
5. Below 0 on a thermometer, the numbers are _____.

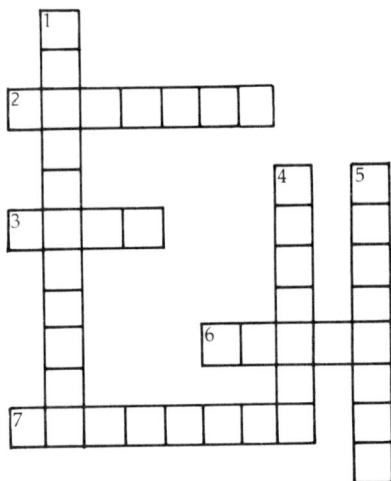

Answers:

Across

2. degrees
3. body
6. point
7. freezing

Down

1. thermometer
4. boiling
5. negative

Temperature—Making Ice Cream

This activity is excellent as a culminating event at the end of the metric unit. At least a forty to fifty minute period should be planned which will include some clean-up time which will be helpful to the teacher.

Among the materials listed are 30 mL test tubes which must be new so that there is no chance of chemical residue. The cost of new test tubes is not great, especially if you consider that they can be used over and over in the ice cream-making activity in future years. Thermometers are also included among the materials. Cheap Celsius thermometers with plastic backs (costing approximately fifty cents each) are excellent. Hopefully you will have access to several of these during the activity. As teachers, you must have a thermometer to use as you go from student to student. The fact that the temperature of the solution goes below the freezing point of water will certainly prompt some questions by inquisitive learners. You might wish to lead a discussion of the use of salt on roads to thaw the ice.

MATERIALS: 30 mL test tubes, Celsius thermometers, styrofoam cups, coffee stirrers, crushed ice, thawing salt, water, milk, vanilla, and sugar.

NUMBER OF PARTICIPANTS: Entire class.

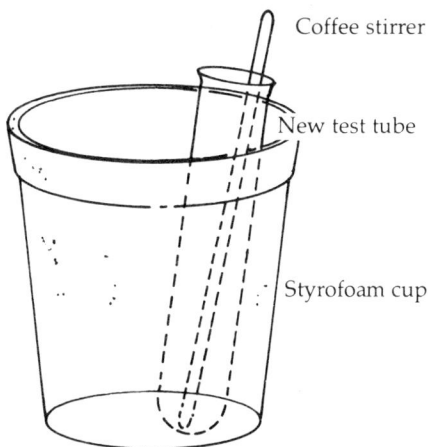

DIRECTIONS: Ingredients: 450 mL of milk, 1 mL of vanilla
45 g of sugar

Prepare the freezing container and ice-salt mixture as follows: Fill a styrofoam cup ¾ full of crushed ice. Add 20 g of thawing salt and enough water to make a solution. Remember to leave enough space in the cup to insert a test tube and a thermometer.

Mix the ice cream ingredients thoroughly before pouring 15 mL of the mixture into each test tube. Place the test tube in the styrofoam cup of ice solution. Place a coffee stirrer or plastic drinking straw in the tube to use in stirring. Occasionally move the test tube around to keep the freezing solution mixed. The coffee stirrer should constantly be moved up and down and around to facilitate the freezing of the ice cream mixture. If the cheap Celsius thermometers are available, place one in the ice-salt solution to check the temperature of the solution. You should find that the temperature of the solution is at least –5°C. The ice cream will freeze at this temperature. Some ice-salt solutions will reach as low as –9°C.

The ice cream is ready to eat when it will not fall out of the tube when held upside down. With the tube in this position, clasp with one hand the portion of the tube containing the frozen ice cream. The warmth of the hand will thaw the edge of the ice cream enough so that it will slip from the tube but retain the form of the test tube. Hold the open hand under the ice cream while eating it.

While the freezing process is underway, the teacher should move through the class checking on progress. It works best to have a thermometer to place in the ice-salt solution of a person who seems to be having difficulty getting the ice cream to freeze. If you find a solution that is only 0°C to –2°C, you can suggest more ice, more salt or a good stirring of the solution. It is not always clear why one solution will be very cold and another not so cold. It is important for the teacher to identify persons having difficulty as it is disappointing to have the ice cream turn out a failure.

APPENDIX A

Principal Units of the Metric System

LENGTH

10 mm	= 1 cm
1000 mm	= 1 m
100 cm	= 1 m
1000 m	= 1 km

WEIGHT

1000 mg	= 1 g
1000 g	= 1 kg
1000 kg	= 1 t

CAPACITY

1000 mL	= 1 L

AREA

100 mm^2	= 1 cm^2
10 000 cm^2	= 1 m^2
100 m^2	= 1 are
1 dm^2	= 1 are
100 ares	= 1 ha
100 ha	= 1 km^2

VOLUME

1 dm^3	= 1000 cm^3
1 m^3	= 1000 dm^3

APPENDIX B*

Comparison of Metric and English Units

LENGTH

1 in = 25.4 mm	1 mm = 0.039 370 in
1 in = 2.54 cm	1 cm = 0.393 701 in
1 ft = 30.48 cm	1 cm = 0.032 808 ft
1 ft = 0.3048 m	1 m = 3.280 840 ft
1 yd = 91.44 cm	1 cm = 0.010 936 yd
1 yd = 0.9144 m	1 m = 1.093 613 yd
1 mi = 1609.344 m	1 m = 0.000 621 mi
1 mi = 1.609 344 km	1 km = 0.621 371 mi

WEIGHT

1 c	= 200 mg	1 mg = 0.005 c
1 c	= 0.2 g	1 g = 5 c
1 oz	= 28 349.523 mg	1 mg = 0.000 035 oz
1 oz	= 28.349 523 g	1 g = 0.035 274 oz
1 oz	= 0.028 349 kg	1 kg = 35.273 962 oz
1 lb	= 453.592 37 g	1 g = 0.002 205 lb
1 lb	= 0.453 592 kg	1 kg = 2.204 603 lb
1 ton (2000 lb) = 907.184 74 kg		1 kg = 0.001 102 ton
1 ton (2000 lb) = 0.907 185 t		1 t = 1.102 311 ton

AREA

1 in² = 6.4516 cm²	1 cm² = 0.155 000 in²
1 ft² = 929.0304 cm²	1 cm² = 0.001 076 ft²
1 ft² = 9.290 304 dm²	1 dm² = 0.107 639 ft²
1 yd² = 0.836 127 m²	1 m² = 1.195 990 yd²
1 acre = 4046.8564 m²	1 m² = 0.000 247 acre
1 acre = 0.404 686 ha	1 ha = 2.471 054 acres

*These comparisons are for information only; they are not intended for use in instructing students.

CAPACITY

1 fl oz = 29.573 527 mL	1 mL = 0.033 813 fl oz
1 fl oz = 0.029 573 L	1 L = 33.814 97 fl oz
1 pt = 473.163 mL	1 mL = 0.002 113 pt
1 pt = 0.473 163 L	1 L = 2.113 436 pt
1 qt = 946.326 mL	1 mL = 0.001 057 qt
1 qt = 0.946 326 L	1 L = 1.056 718 qt
1 gal = 3.785 306 L	1 L = 0.264 179 gal

VOLUME

$1 \text{ in}^3 = 16\ 387.064 \text{ mm}^3$	$1 \text{ mm}^3 = 0.000\ 061 \text{ in}^3$
$1 \text{ in}^3 = 16.387\ 064 \text{ cm}^3$	$1 \text{ cm}^3 = 0.061\ 024 \text{ in}^3$
$1 \text{ ft}^3 = 28\ 316.847 \text{ cm}^3$	$1 \text{ cm}^3 = 0.000\ 035 \text{ ft}^3$
$1 \text{ ft}^3 = 28.316\ 847 \text{ dm}^3$	$1 \text{ dm}^3 = 0.035\ 316 \text{ ft}^3$
$1 \text{ yd}^3 = 764.555 \text{ dm}^3$	$1 \text{ dm}^3 = 0.001\ 308 \text{ yd}^3$
$1 \text{ yd}^3 = 0.764\ 555 \text{ m}^3$	$1 \text{ m}^3 = 1.307\ 951 \text{ yd}^3$

APPENDIX C*

Conversion Tables**

LENGTH

To Change:	Multiply by:
miles to kilometers	1.6
miles to meters	1609.3
yards to meters	0.9
inches to centimeters	2.54
inches to millimeters	25.4
feet to centimeters	30.5
kilometers to miles	0.62
meters to yards	1.09
meters to inches	39.4
centimeters to inches	0.39
millimeters to inches	0.04

WEIGHT

To Change:	Multiply by:
tons to metric tons	0.9
pounds to kilograms	0.45
pounds to grams	453.6
ounces to grams	28.4
metric ton to tons	1.1
kilograms to pounds	2.2
grams to pounds	0.002
grams to ounces	0.035

*Conversion tables are given for information only; they are not intended for use in instructing students.
**The conversion factors have been rounded to facilitate in computations.

AREA

To Change:	Multiply by:
square inches to square centimeters	6.45
square feet to square meters	0.093
square yards to square meters	0.836
acres to hectares	0.4
square miles to square kilometers	2.6
square centimeters to square inches	0.155
square meters to square feet	10.8
square meters to square yards	1.2
hectare to acres	2.5
square kilometers to square miles	0.386

VOLUME

To Change:	Multiply by:
cubic inches to cubic centimeters	16.4
cubic feet to cubic meters	0.03
cubic yards to cubic meters	0.76
cubic centimeters to cubic inches	0.06
cubic meters to cubic feet	35.0
cubic meters to cubic yards	1.3

CAPACITY

To Change:	Multiply by:
teaspoons to milliliters	5
tablespoons to milliliters	15
fluid ounces to milliliters	29.6
pints to milliliters	473
pints to liters	0.47
quarts to liters	0.95
gallons to liters	3.8
bushels to liters (dry)	35.2
milliliters to fluid ounces	0.034
liters to pints	2.1
liters to quarts	1.06
liters to gallons	0.26
liters to bushels (dry)	0.03

TEMPERATURE

To change C to F, multiply C by 1.8 and add 32.

To change F to C, subtract 32 from F and divide by 1.8.

Practical Comparisons

1. The meter is a little longer than a yard.
2. A kilometer is about 0.6 of a mile.
3. The kilogram is a little heavier than 2 pounds.
4. The metric ton is a little more than a ton.
5. The liter is a little more than a quart.
6. Human body temperature is 37°C.
7. A hectare is about 2.5 acres.

Notes

[1]"Going Metric: A Lot Easier and Cheaper Than Expected." *U.S. News & World Report,* 81:114–15; July 1976.

[2]W. Knowles Middleton, *A History of the Thermometer and Its Use in Meteorology* (Baltimore: Johns Hopkins University Press, 1966).

[3]Richard Elwell, "The Inevitable Metric Advance," *American Education,* 12:6–9; December 1976.

[4]The shortened forms of metric units are symbols, not abbreviations; therefore, no period should be used after the shortened form.

[5]U. S. Congress, *Federal Register* (Washington D.C.: Government Printing Office, December 10, 1976).

[6]American National Metric Council, *Metric Guide for Educational Materials* (Washington D.C.: American National Metric Council, 1977).

[7]Since commas are used as decimal markers in many countries, they should not be used to separate groups of digits. The Educational Materials Sector Committee of the American National Metric Council recommends the gradual adoption of these practices for writing numerals and complete adoption in all publications with a copyright date of 1982 or later.

[8]American National Metric Council, *Metric Guide.*

[9]Richard W. Copeland, *How Children Learn Mathematics: Teaching Implications of Piaget's Research,* 2nd ed. (New York: Macmillan Co., 1974).

[10]Irv King and Nancy Whitman, "Going Metric in Hawaii," *The Arithmetic Teacher,* 20:258–60; April 1973.

SELECTED RESEARCH REFERENCES

Alexander, F. D. "The Metric System—Let's Emphasize Its Use in Mathematics." *Arithmetic Teacher* 20:395–96; May 1973.

Audiovisual Instruction. "Who's Who in Metrics—A Selected Listing." *Audiovisual Instruction* 20:11; February 1975.

Barbrow, L. E. "Metrication in Business and Industry." *Business Education Forum* 28:8–9; December 1973.

Batcher, Olive M., and Young, L. A. "Metrication and the Home Economist." *Journal of Home Economics* 66:28–31; February 1974.

Bitter, G. "Metric Corner." See issues of *Teacher,* September 1976.

Blondell, Beverley. "Going Metric; Office of Education's Metric Education Program." *American Education* 12:34–35; December 1976.

Bolster, L. Carey. "Activities: Centimeter and Millimeter Measurements." *Mathematics Teacher* 67:623–26; November 1974.

Bright, George W. "Bilingualism in Measurement: The Coming of the Metric System." *Arithmetic Teacher* 20:397–99; May 1973.

———. "Metrics, Students and You!" *Instructor* 83:60–65; October 1973.

———, and Jones, Carolann. "Teaching Children to Think Metric." *Today's Education* 62:16–19; April 1973.

Brooks, Mattie J. "One-Meter Dash." *Arithmetic Teacher* 24:327–28; April 1977.

Brundage, D. "Some Dos and Don'ts in Metric; American Institute for Research Reports Going Metric and Metric Inservice Teacher Training." *American Education* 12:10–13; December 1976.

Bruni, James V., and Silverman, H. J. "An Introduction to Weight Measurement." *Arithmetic Teacher* 23:4–10; January 1976.

———, and ———. "Organizing A Metric Center in Your Classroom." *Arithmetic Teacher* 23:80–87; February 1976.

Burton, Grace M. "A Potpourri of Metric Activities." *Elementary School Journal* 76:201–7; January 1976.

Caravella, Joseph R. "Metrication Activities in Education." *Business Education Forum* 28:14–16; December 1973.

Cathcart, W. George. "Metric Measurement: Important Curricular Considerations." *Arithmetic Teacher* 24:158–60; February 1977.

Chalupsky, Albert B. "Metric Instructional Materials: Learning from Other Countries." *Audiovisual Instruction* 20:8–10; February 1975.

———, and Crawford, J. J. "Preparing the Educator to Go Metric." *Phi Delta Kappan* 57:262–65; December 1975.

Chipley, Donald R. and Trueblood, C. R. "Monitoring the Move to Metrication: A Research Strategy and Survey Summary." *School Science and Mathematics* 76:703–11; December 1976.

Choate, Stuart A., comp. "A Metric Bibliography." *Mathematics Teacher* 67:586–87; November 1974.

Clason, Robert G. "1866; When the United States Accepted the Metric System." *Arithmetic Teacher* 24:56–62; January 1977.

Cochran, Marianna Z. "Cooking with Grams." *Journal of Home Economics* 67:31–34; January 1975.

Cortright, Richard W. "Conversion to the Metric System: Roles of the National Education Association." *Audiovisual Instruction* 20:17; February 1975.

Drake, Paula. "Hello Metrics!" *Teacher* 91:46–50; October 1974.

Dubisch, Roy. "Some Comments on Teaching the Metric System." *Arithmetic Teacher* 23:106–7; February 1976.

Elwell, Richard. "Inevitable Metric Advance." *American Education* 12:6–9; December 1976.

Epstein, Susan L. "Weighing Ideas." *Arithmetic Teacher* 24:293–97; April 1977.

Firl, Donald H. "The Move to Metric: Some Considerations." *Mathematics Teacher* 67:581–84; November 1974.

Fisher, Ron. "Metric Is Here; So Let's Get on With It." *Arithmetic Teacher* 20:400–402; May 1973.

Freeman, William W. K. "Think Metric About Weather." *Arithmetic Teacher* 22:378–81; May 1975.

Grzesiak, Katherine A. "America and the Metric System: Present Perspectives." *Elementary School Journal* 76:195–200; January 1976.

Hallerberg, Arthur E. "The Metric System: Past, Present, Future?" *Arithmetic Teacher* 20:247–55; April 1973.

Hatfield, Thomas A. "Art Program and the Metric System." *School Arts* 76:60–61; March 1977.

Hawkins, Vincent J. "Teaching the Metric System as Part of Compulsory Conversion in the United States." *Arithmetic Teacher* 20:390–94; May 1973.

———. "Some Changes in Shop Mathematics Due to Metrication." *Mathematics Teacher* 67:601–3; November 1974.

Helgren, Fred J. "Schools are Going Metric." *Arithmetic Teacher* 20:265–67; April 1973.

Hoffman, Joseph R. and Henry, Boyd. "Weight Watcher's Test." *Mathematics Teacher* 70:344–46; April 1977.

Hugelman, Opal. "Me and Metric." *Audiovisual Instruction* 20:27; February 1975.

Jones, Philip G. "Metrics: Your Schools Will Be Teaching It and You'll Be Living It—Very, Very Soon." *American School Board Journal* 160:21–27; July 1973.

King, Irv, and Whitman, Nancy. "Going Metric in Hawaii." *Arithmetic Teacher* 20:258–60; April 1973.

Lindquist, Mary Montgomery and Dana, Marcia E. "Let's Do It: The Neglected Decimeter." *Arithmetic Teacher* 25:10–17; October 1977.

Marcuccio, Phyllis. "Metric Education: Trends and Recommendations." *Science and Children* 14:7–10; February 1977.

Meiring, Steven P. "Metric Measurement and Instructional Television." *Arithmetic Teacher* 25:42–44; October 1977.

Morehouse, Thomas and Schoonmaker, Edwin. "Metric Month at Taft Middle School." *Phi Delta Kappan* 57:265; December 1975.

Moss, Jeanette K. "Teaching Aids: Tooling Up for the Metric Changeover." *Teacher* 91:90–97; March 1974.

Muellen, T. K. "Metric in Maryland." *Educational Leadership* 31:435–37; February 1974.

National Council of Teachers of Mathematics, Metric Implementation Committee. "Metric: Not If, But How." *Arithmetic Teacher* 21:366–69; May 1974.

——, Metric Implementation Committee. "Metric Competency Goals." *Mathematics Teacher* 69:90–91; January 1976.

Nation's Schools. "Students Learn to Live with Liters and Meters." *Nation's Schools* 93:24–25; April 1974.

Odom, Jeffrey V. "The Current Status of Metric Conversion." *Business Education Forum* 28:5–7; December 1973.

——. "The Effect of Metrication on the Consumer." *Business Education Forum* 28:10, 12–13; December 1973.

Pottinger, Barbara. "Measuring, Discovering, and Estimating the Metric Way." *Arithmetic Teacher* 22:372–77; May 1975.

Rieck, William A. "Conversion to the Metric System: When and How?" *Clearing House* 50:80–82; October 1976.

Ropa, Adrienne. "Roll Out the Meters." *Instructor* 86:78–79; May 1977.

Sak, Theresa. "Metric Quiz; Test Your SI Savvy." *Instructor* 86:58–59; October 1976.

Sengstock, Wayne L. and Wyatt, K. E. "Meters, Liters, and Grams." *Teaching Exceptional Children* 9:58–65; Winter 1976.

Straub, Sylvia. "Metric Mastery Through TV." *Teacher* 94:29–30; February 1977.

Stuart, Kristine. "Metric Made Fun: An Individualized Approach." *Science and Children* 14:11–13; February 1977.

Thompson, Thomas E. "Ten Commonly Asked Questions by Teachers about Metric Education." *Science and Children* 14:14–16; February 1977.

Vervoort, Gerardus. "Inching Our Way Toward the Metric System." *Arithmetic Teacher* 20:275–79; April 1973.

Viets, Lottie. "Experiences for Metric Missionaries." *Arithmetic Teacher* 20:269–73; April 1973.

Webb, Leland F. "Measuring Science and History; Time-line Problem." *Arithmetic Teacher* 24:115–16; February 1977.

West, Tommie A. "Teaching Metrics to Beginners." *Today's Education* 63:80–82; November/December 1974.

Wilderman, Ann. "Metric Generation." *Teacher* 94:75–80; March 1977.

Wilderman, Ann M., and Krulik, Stephen. "Metric Resources: What the Teacher Will Want." *Audiovisual Instruction* 20:29–30; February 1975.

Williams, David E. and Wolfson, Brian. "Play Metric; Games to Help Kids Think Metric." *Instructor* 86:62–63; April 1977.

Williams, Elizabeth. "Metrication in Britain." *Arithmetic Teacher* 20:261–64; April 1973.